Organic Semiconductors for Optoelectronics

ORGANIC SEMICONDUCTORS FOR OPTOELECTRONICS

S.N. Shukla

ARCLER
PRESS

www.arclerpress.com

Organic Semiconductors for Optoelectronics

S.N. Shukla

Arcler Press
224 Shoreacres Road
Burlington, ON L7L 2H2
Canada
www.arclerpress.com
Email: orders@arclereducation.com

ISBN: 978-1-77469-530-2 (Hardcover)

ABOUT THE AUTHOR

SachchidaNand Shukla, is presently working as Professor, Department of Physics & Electronics, Dr. RamManohar Lohia Avadh University, Ayodhya, UP, India. He did his Masters in Physics (Electronics) in 1988 and Ph.D. in 1992 from the same university.

Dr. Shukla holds 27 years experience of teaching M.Sc. (Physics) and M.Sc. (Electronics) students and 3 years experience of teaching B.Tech., MCA and B.Sc. (Electronics) students. He has published 85 research papers in peer-reviewed/ indexed journals of International/National repute & conference proceedings and 02 books. In his supervision 14 research scholars have been awarded Ph.D. degree.

In addition to it he is the recipient of Best Scientist National Award (2018) of IRDP Group of Journals, Chennai and Maatee Ratan Samman (2017). He has also been selected as Fellow of IACSIT (International Association of Computer Science and Information Technology, Singapore) and Associate Fellow of IAPS (International Academy of Physical Sciences) in 2018. In view of Dr. Shukla's academic achievements his employer institution, Dr. RamManohar Lohia Avadh University, has conferred upon him the 'Certificate of Appreciation' in 2018.

Besides having a wide exposure to various key positions of University administration like Pro Vice Chancellor, Registrar, Director College Development Council (CDC), Coordinator UGC and RUSA, Head of Physics Department etc, Dr. Shukla has membership of 08 academic bodies of international repute. To name a few are ISCA (The Indian Science Congress Association, Kolkata, India), IETE (The Institution of Electronics and Telecommunication Engineers, New Delhi, India), NASI (The National Academy of Sciences, India, Allahabad, India), IAPS (International Academy of Physical Sciences, Allahabad, India) and SCIEI (Science and Engineering Institute, Hong Kong, SAR of China).

In addition, he is also gracing the Editorial Boards of 04 international journals IJAREEIE (International Journal of Advanced Research in Electrical, Electronics and Instrumentation Engineering), APM (Applied Physics & Mathematics), JUSPS (Journal of Ultra Scientist of Physical Sciences) and IRJ (International Researcher's Journal) and the board of reviewer of IRJECE (International Journal of Electronics & Communication Engineering).

TABLE OF CONTENTS

LIST OF FIGURES

LIST OF TABLES

LIST OF ABBREVIATIONS

1D	one-dimensional
AOs	atomic orbitals
APDs	avalanche photodiodes
ASE	amplification spontaneous emission
B	boron
BCB	benzocyclobutene
CCD	charge-coupled device
CMOS	complementary metal-oxide semiconductor
CRI	color rendering index
CSS	charge-separated states
CT	charge-transfer
CTS	charge transport state
CVD	chemical vapor deposition
D	detective
DR1	disperse red 1
DTBTT	dithienobenzotrithiophen
EA	electroabsorption
EL	electroluminescence
ENDOR	electron-nuclear double resonance
EQE	external quantum efficiency
ESR	electron spin resonance
ETL	electron-transport layer
FRET	Förster resonance
FWHM	full-width-at-half-maximum
HOMO	highest occupied molecular orbital
HT/ET	hole and electron-transporting
HTL	hole-transport layer
HV	high vacuum
IQE	internal quantum efficiency
ISC	intersystem crossing

ITO	indium tin oxide
LANs	local area networks
LCAO	linear combination of atomic orbitals
Ld	diffusion length
LDR	linear dynamic range
LEDs	light-emitting diodes
LUMO	lowest unoccupied molecular orbital
LV	lower vacuum
MC	merocyanine
MWCNTs	multi-walled carbon nanotubes
NEP	near-infrared
O4PC	oxazine 4 perchlorate
ODMR	optically detected magnetic resonance
OE	organic electronics
OFETs	organic field effect transistors
OLEDs	organic light-emitting diodes
OMC	organic molecular crystals
OPDs	organic photodetectors
OPV	organic photovoltaic
OSCs	organic semiconductors
OSOAs	organic semiconductors optical amplifiers
OTFTs	organic thin-film transistors
P	phosphorus
P3HT	poly(3-hexylthiophene)
PDS	photothermal deflection spectroscopy
PF	polyfluorenes
PL	photoluminescence
PLQY	photoluminescence quantum yield
PMMA	polymethylmethacrylate
POF	polyolefin fibers
PPP	Pariser-Parr-Pople
PPP	poly para phenylene
PPV	poly(phenylene vinylene)
PTCD	perylene tetracarboxylic diimide

PTCDA	perylene tetracarboxylic dianhydride
Q	quinoline
QWs	quantum wells
RB	rhodamine B
S-AS	soliton/anti-soliton
Si	silicon
SNR	signal-to-noise ratios
SP	surface plasmon
SPP	surface plasmon polariton
SSH	Su-Schrieffer-Heeger
SWCNTs	single-walled CNTs
TCR	temperature coefficient of resistance
TPA	triphenylamine
UHV	ultra-high vacuum
UV	ultraviolet
VLCs	visible light communications
Voc	open circuit voltage
VSL	varying stripes length
WDM	wavelength-division multiplexed
XPS	X-ray photoelectron spectroscopy
ZnPc	zinc phthalocyanine

PREFACE

Organic semiconductors' (OSC) photoconductive and semiconducting capabilities were first described in 1906 and 1950, respectively, and since then, fundamental research has been carried out constantly. As photoreceptors for electrophotography, molecularly dispersed polymers—insulating polymers that contain hole transport molecules—were first commercialized in 1980. This organic photoreceptor was produced using a coating method, which reduced the price of the photoreceptor. Organic solar cells and organic light-emitting diodes (OLEDs) were first published in 1987 and 1989, respectively. These gadgets demonstrated the possibilities of organic devices at the time and were very effective. High-definition OLED TVs and OLED lighting are currently available. OLEDs were first utilized as an automotive display in 1997. Future flexible screens, biosensors, and other devices that would be impossible to make with standard inorganic semiconductors are predicted to be successfully made with organic semiconductors. Without a thorough grasp of the optoelectronic characteristics of organic semiconductors and how these characteristics affect the overall performance of the device, it will be impossible to create future organic devices.

The goal is to have a single volume that includes current advancements in the subject and covers everything from the basics to applications. This book discusses some recent advancements in the study of organic semiconductors' optoelectronic properties as well as summarizes the fundamental ideas. In the area of electronic and optoelectronic organic materials, examples, and applications are covered. This book, which comprises 8 chapters contributed by experienced and renowned scientists on various areas of optoelectronic properties of organic semiconducting materials, makes an effort to address both experimental and theoretical breakthroughs in each topic. The majority of chapters are presented to be fairly independent with little cross-referencing; nevertheless, chapters with complementary material are grouped together to make cross-referencing easier for the reader.

There are a total of eight distinct chapters in the book. In Chapter 1, the reader receives an introduction to organic semiconductors. In Chapter 2, the structure and characteristics of organic semiconductors are covered in great detail. In Chapter 3, the topic of organic semiconductors for device applications is explored in great detail.

The book's readers are given an introduction to optoelectronic devices in Chapter 4. Chapter 5 focuses heavily on organic semiconductors for optical applications. In Chapter 6, a breakdown and illustration of the organic semiconductors for photodetectors are provided. Chapter 7 also covers organic semiconductors for visible light communication. Light emission is covered in Chapter 8, "Plasmonics for Light-Emitting and Photovoltaic Devices."

The many distinct facets of organic semiconductors are comprehensively covered in this book. Even a reader with no prior knowledge should have no issue comprehending the basic concepts behind organic semiconductors after reading this book.

—Author

CHAPTER 1

INTRODUCTION TO ORGANIC SEMICONDUCTORS

CONTENTS

1.1. INTRODUCTION

Organic semiconductors (OSC) are a significant class of optical and electrical materials which are the subject of a significant amount of research. This is because it is possible to tune their characteristics by optimizing their chemical composition and it is simple to fabricate organic semiconductor materials into adaptable films as well as devices. Even though the majority of work has been spent on building displays and illumination from all these substances, the unique features that they possess also make them interesting in visible light communications (VLCs). This chapter discusses the primary applications which have been investigated and describe how the qualities of these materials make them appropriate for VLC (Newman et al., 2004; Brütting, 2005). Just on the transmitting end, recording white VLC transmission has been accomplished by using inorganic semiconductors as color converters. In addition, direct modulation with organically light-emitting diodes (LEDs) also was conceivable and might be of relevance for display-to-display communications. On the receiving end, solar cells may be utilized to concurrently capture data over the network, while fluorescence antennas provide quick and sensitive receivers with such a vast visual field. Both of these features are possible thanks to the usage of fluorescence antennas (Takimiya et al., 2011; Myers and Xue, 2012).

OSCs are getting so much attention those same days even though they have a lot of appealing characteristics that established them apart from their traditional inorganic equivalents. Some of these characteristics include being extremely lightweight, having a low cost of production, being able to be processed at such a low temperature, having structural flexibility, and having abundant accessibility. The recent high-profile commercialization of complex organic light-emitting diodes (OLEDs) with elevated smartphone displays provides further evidence that organic molecules have good potential for usage in a wide variety of other electrically and optoelectric technologies (Knupfer, 2003; Köhler and Bässler, 2009). For instance, the use of OSCs in the production of solar cells, transistors, photodiodes, and laser beams is capable of achieving very high levels of both feasibility and efficiency. One other advantageous quality of OSCs was that they're being processed utilizing simple solution processing methods, such as ink-jet printing as well as film fabrication. This characteristic makes the production of electrical devices much less difficult and more cost-effective. In the speech that he gave for the Nobel Prize, Professor Heeger stated, "I'm certain like we're on the eve of a revolutionary in 'Plastic Electronics.'" We anticipate that significant advancements in the area of OSCs will bring to the creation of organic devices that are less costly, more practical, and more intelligent as time goes on (Yamashita, 2009; Nitti et al., 2021).

Organic molecules are compounds that consist mostly of hydrocarbon molecules; for a significant amount of time, it was believed that these were the only kind of substances that could function as electrical insulators. The finding that organic molecules may exhibit electrical properties can be dated back to 1950; as just a result, an exciting new area of study has been made possible as a result of this discovery. In 1963, Pope et al. published a study that demonstrated electroluminescence (EL) coming from single crystals of anthracene. After exposing polyacetylene to the vapors of iodine, chlorine, or bromine, in the 1970s, Shirakawa, Heeger, and MacDiarmid discovered that the permeability of polyacetylene might be greatly increased, reaching a level that was comparable to those of a typical metal (Dong et al., 2010; Jacob, 2014). This discovery was made possible by the discovery that polyacetylene might be exposed to vapors of these elements. The discovery of such high electrical conductivity in a material traditionally thought of as an insulator attracted a great deal of interest and fueled the area of organic electronics' (OEs) quick expansion. Heeger, MacDiarmid, and Shirakawa were presented with the Nobel Prize for Chemistry during the year 2000 for their contributions to "the detection and expansion of leading polymers" (Krause et al., 2008; Lin et al., 2012).

The technology behind OLEDs, has been receiving a lot of attention recently among the many organically electronic gadgets that are available today. In beginning, the EL that was seen coming from organically single crystals has not been very good at conserving energy and needed a significant amount of power to operate the devices. In 1987, Tang and Van Slyke produced light-emitting technologies based on unstructured depositing a thin layer. Using an electric biased of less than 10 V, they were able to attain a quantum yield of 1% in their devices. Professor Friend at Cavendish Laboratory also disclosed in 1990 the development of OLEDs that used conjugated polymers. Since that time, key technologies, such as advanced materials, and device structures, including device engineering techniques, have grown fast, gradually forming new industries. These developments include advanced materials, device structures, including technology engineering techniques (Brédas and Marder, 2016; Bronstein et al., 2020).

As chemically synthesized gadgets have progressed from scientific ideas within research labs to application areas just on the proper market, there seems to be a necessity to comprehend the distinctive properties of natural semiconductor compounds (OSCs) and evaluate the possibilities that even this new generation of OEs might offer. This is because there is a necessity to comprehend the special features of OSCs. In just this chapter, we talk about the electrical structures of OSCs, as well as numerous basic charge storage situations that may occur in OSCs (Nakano et al., 2015; Hofmann et al., 2019). In addition to that, we examine the idea of doping in OSCs. Our discussion will come to a close with a concise summary of several fascinating technological applications. These applications include electrophotographic electronics, and thin film transistors, including light-emitting transistors (Schwarze et al., 2016).

OSC are a specialized category of polymeric products that have their own set of distinctive electrical and optical characteristics. They are especially intriguing because they combine unique semiconductors' electrical and optical capabilities with cheap production and the ability to tune aspects related to organic substances. This makes them highly attractive. In contrast to semiconductor materials, OSC wouldn't need a highly organized crystalline structure to display the electronic and optical capabilities of semiconductors. These characteristics are mostly attributable to the chemical composition of semiconductor materials. Because of this, the manufacture of their devices is made much easier—for instance, via deposition from solution—which is anticipated to result in a reduction in the overall cost of the devices. In addition to this, it makes it possible to fabricate many devices simultaneously, such as the red, green, and blue organically light-emitting diode (OLEDs) that are used in the displays of a smartphone or televisions (Okamoto et al., 2012; Torabi et al., 2015).

OSC get their optical and electronic capabilities through their chemical composition, which is called conjugated and consists of alternate carbon-carbon double bonds among nearby carbon atoms. This gives the OSC unique electrical and optical properties. This ultimately results in the overlapping of electronically electron shells as well as the delocalization of electrons (Figure 1.1). The carbon atoms in linked compounds have a sp^2-hybridized configuration. Three of a molecule's four valence electrons are located in orbitals that would be in the planes of the molecule. These orbitals create–bonds, that are responsible for the molecule's cohesiveness. The *p* orbital that contains the fourth valence electron may be found extending further out on the planes that the molecule occupies (Shen et al., 2004). The intersection among the 2pz orbitals consequences in the creation of π-bonds and (when there are numerous atoms) π and π* bands. The delocalization of electrons throughout the length of the molecule, together with the discovery that there are two of these electrons for every repeat unit, results in such a band that is full and exhibits semiconductors' optical and electrical characteristics. This indicates that they are suitable for use in the production of a variety of semi-conducting technologies, such as laser beams, LEDs, solar panels, and transistors. The natural semiconductor materials that are most often utilized originate from one of three categories: small particles, polymers, or dendrimers. Most are soluble, which makes depositing from solutions a fairly straightforward process; others are evaporated into thin films for use in electronic devices and then deposited (Meng et al., 2009; Ma et al., 2009).

Visible range transmission is a relatively recent area during which OSC were beginning to find many uses (VLCs). Even though the beam is used for information exchange since antiquity, optical fibers have only dominated communication systems in the last 30 years, and also more recently, there's been a growing interest in free storage VLC (Henson et al., 2012). This is even though the light is being used for information exchange since antiquity. Light has several attractive qualities for transmitting data, including hundreds of terahertz license-free bandwidth, (ii) simple front-end devices, (iii) no intervention with sensitive equipment, and (iv) the possibility for assimilation through into existing lighting facilities. Light also was recognized as Li-Fi, which is limited for light fidelity. Light is also known as Li-Fi (short for light fidelity) (Jiang et al., 2013; Abate et al., 2014).

Color conversions, OLEDs, photodiodes (with both organic photovoltaics (OPVs) and organic photodiodes (OPDs)), and optically antennas are the four primary types of devices that are based primarily on semiconductor materials in VLC applications. Such devices take advantage of the unique characteristics that OSC provide in a variety of different ways. For instance, the widespread delocalization of electrons results in a very significant rate of absorption and, as a direct consequence of this, a strong radiation rate of emission. When used to color converters and optical antennas, this results in a rapid on-off reaction. When it comes to optimizing the absorbing of OPVs and OPDs, as well as the emissions of OLEDs, having the ability to modify the energy shortage by changing its chemical composition may be of great assistance (Sergeyev et al., 2007; He et al., 2016).

Outcomes in terms of the bandwidth that can be achieved and the data rates that can be achieved are provided throughout this whole chapter. In this analysis, the term "bandwidth" refers to the rate with which the electromagnetic power output of both the detectors drops to a level of–3 dB, which would be equivalent to 50% of its value when referenced to DC. Data rates also are given, even though they are impacted by a wide variety of parameters including the encoding technique that is used and the possible SNR. When it is within our means to do so, we disclose the most important experimental information that is connected to the stated data rates (Sun et al., 2015).

1.2. ELECTRONIC STRUCTURES

According to the molecular mass of various organic materials, OSCs are often divided into two primary organizations: linked polymers (Figure 1.1) and tiny molecules (Figure 1.2). The repeating structural units (monomers) that makeup polymers are joined by hydrogen bonding. They are manufactured with a variety of molecular masses because the amount of repeating units is often difficult to manage exactly. Small molecules, but on the other side, have exact, much lower molecular weight. Those two OSC groups also differ in terms of how they are manufactured. The majority of conductive polymers may be handled utilizing different solution-processing techniques since they are soluble in nonpolar solvents (Pecile et al., 1989). Molecules, on the other hand, are often formed using heating evaporation in a high vacuum (HV), a depositing technique that gives the organic films' thicknesses, shape, and other characteristics excellent control. As a result, it is simple to produce high-quality, extremely thin (even reduced to just a few nanometers) films of tiny molecules. Small compounds' chemical structures may be modified to make them accessible in polar solvents, which makes solution manufacturing procedures easier to utilize. Pentacene, for instance, can only ever be deposited by heat evaporate (Figure 1.2(d)), but 6,13-bis[(triisopropylsilyl)ethynyl]pentacene (TIPS-pentacene) (Figure 1.2(f)) has covalent bonds that enhance its solubility, make treating it simpler, and retain high crystalline nature (Allard et al., 2008; Turkoglu et al., 2019).

We start with both the atomic orbitals (AOs) of carbon, the most significant component of all organic compounds, to comprehend the electrical structures of OSCs. Six electrons make up a carbon atom, which has the electrical configuration (1s)2(2s)2(2p)2. Several hybridized orbitals, including that of the *sp*, *sp*², as well as *sp*³ hybrids orbitals, may be used to connect it to other atoms (Figure 1.3). Triple *sp*² hybrid orbitals, for instance, are created when ones, as well as two *p* orbital angular momentum, are quantum superpositions; the resonances are spread in the same plane at a 120° angle. Each carbon atom inside the simple model molecule ethylene provides three *sp*² hybrid orbitals for forming bonds with other atoms (Figure 1.4) (Takimiya et al., 2007). A molecular orbital is created when the hydrogen atoms 1s orbiting intersects with one of the *sp*² hybrid orbitals (MO), a so-called s-orbital; the link formed as a consequence is an s-bond.

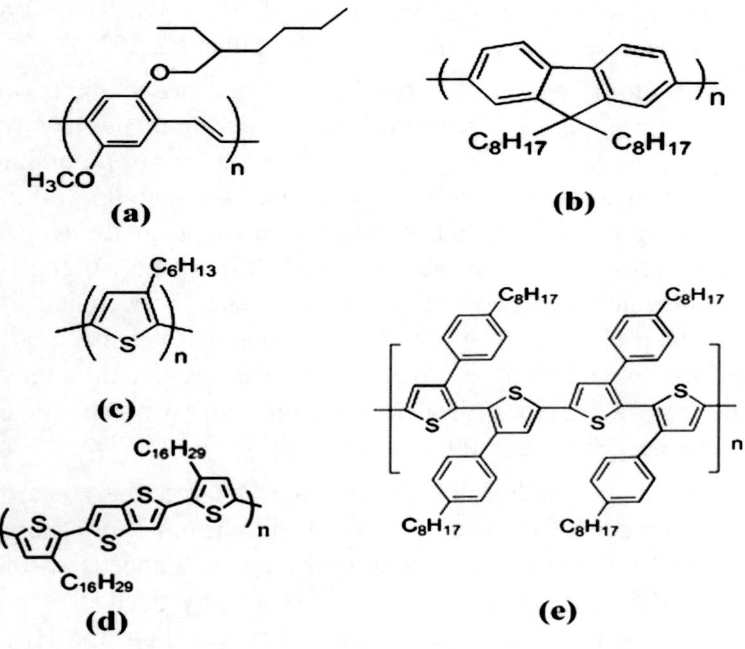

Figure 1.1. Chemical compositions of polymer-based OSCs include (a) poly[1-methoxy-4-(2-ethylhexyloxy)-p-phenylene vinylene (MEH-PPV); (b) poly(9,9 dioctylfluorene) (PFO); (c) poly(3-hexylthiophene-2,5-diyl) (P3HT); (d) poly[2,5-bis (POPT).**(e) poly[3-(4-octylphenyl)thiophene](POPT).**

Source: https://www.researchgate.net/figure/Chemical-structures-of-cyanated-phenylenevinylene-based-polymer-acceptors-a-CN-PPV_fig25_298738133.

Additionally, the 2 carbon atoms are joined together by *sp*² hybrid electron shells. Nevertheless, a *p* atomic orbital that is perpendicular to the planes of the *sp*² hybrid electron shells is still retained for every carbon atom. The second sort of MO known as a p-orbital is created when the two Pz two electrons overlap one another side-by-side with both the electron density dispersed along each internuclear axis. The p-electron mist among the dual carbon particles is delocalized, as you can see. As a result, two bonds—an s-bond and a p-bond—join the carbon atoms, forming what is known as a "double bond" in this configuration (Ando et al., 2005). The overlap of two AOs results in two MOs, a bonding phase, and an anti-bonding state, following constructive and destructive interference, correspondingly, according to the electron wave equation. The power of the anti-bonding phase is greater than the energy of the bonding state even though the interference pattern causes a decrease in the electron density between atoms. The bonding, as well as anti-bonding states of p-orbitals, are referred to as p- and p*-MOs, respectively (Liu et al., 2014; Hu et al., 2021).

6 carbon atoms are linked together in benzene, a bigger molecule (Figure 1.5), employing sp^2 hybrid orbital angular momentum as well. The remaining six Pz AOs similarly combine to produce six p-bonding MOs, which consist of three bonded as well as three anti-bonding orbitals. The bottom bonding and higher anti-bonding states degenerate as a result of interference between the six pz electron shells (Figure 1.5(b)). In a balanced benzene molecule, each of the three linking states is occupied by six pz electrons. The p-electron swarm is delocalized across the six carbon atoms, much as in ethylene. In those other words, a molecular orbitals structure that allows for "conjugation" has electrons that can flow from one carbonyl group to another (Forrest, 2000; Qi et al., 2008).

The highest occupied orbital (MO) is the MO with the greatest energy which electrons occupy (HOMO). The MO with the minimum energy which electrons need not occupy is known as the lowest unoccupied molecular orbital (LUMO). The energy disparity. The energy needed to accelerate the electrons within a molecule is located between both the HOMO and LUMO. For instance, in benzene (Figure 1.5(b), the higher bonding MOs (p2, p3) are the HOMOs and the lower anti-bonding MOs (p×, p×) are the LUMOs. For OSCs, because the electronic activities (such as photon absorption or emission) take place energetically favorably on p-orbitals, and dynamic splitting of p-bonds is often lower than either s-bonds. Similar to this, charges introduced from an organic semiconductor device's working electrode would typically occupy p-Mos (Yamada et al., 2008). In other words, the p-electron system's border orbitals, which have lower and more advantageous energy levels, are where OSCs' electronic functions emerge. In contrast, the s-electron system's electrons are barely stimulated since doing so requires considerably more energy. Additionally, because of the larger degree of p-electron delocalization, the HOMO-LUMO energy gap shrinks as the size of the p-conjugation system grows (Sokolov et al., 2012).

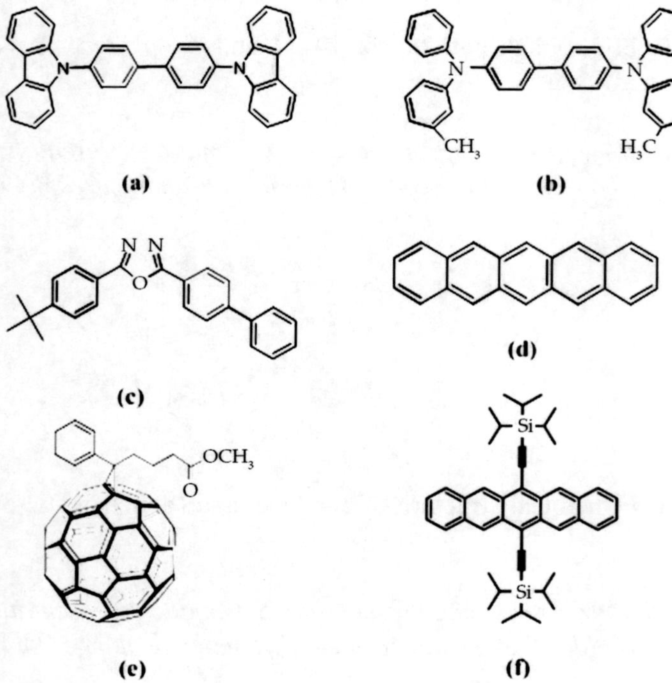

Figure 1.2. Small-molecule semiconductors' chemical structures are as follows: (a) 4,4-N,N0-dicarbazolyl-biphenyl (CBP); (b) N,N0-diphenyl-N,N0-bis (3-methylphenyl) TPD; (c) 5-1,10-biphenyl-4,40-diamine (4-biphenyl) pentacene, [6,6]-phenyl-C61-butyric acid methylester, [4,4-tert-butylphenyl]-1,3,4-oxadiazole (t-PBD), [6,6]-phenyl-C61-butyric acid methylester (PCBM), and [6,13-bis[(triisopropylsilyl)ethynyl]pentacene)).**(d)pentacene,(e)[6,6]-phenyl- C61-butyric acidmethylester(PCBM),and(f)6,13-bis[(triisopropylsilyl) ethynyl]pentacene((TIPS-pentacene)solublepentacene).**

Source: https://pubs.rsc.org/en/content/articlelanding/2010/jm/b926348a/unauth.

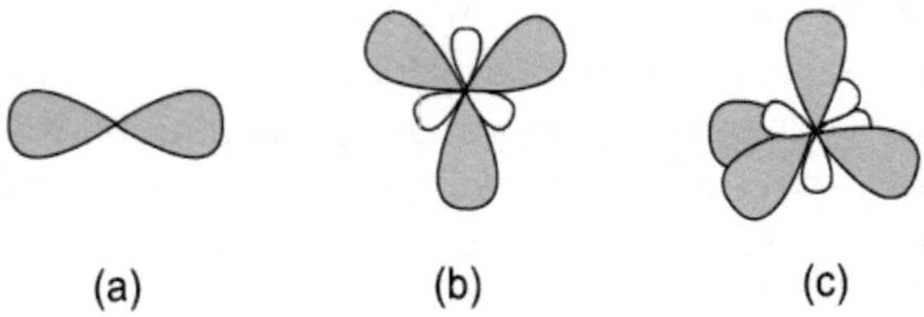

Figure 1.3. Hybrid orbitals with (a) sp; (b) sp^2; and (c) sp^3 are shown schematically.

Source: https://www.researchgate.net/figure/Schematic-representation-of-a-p-orbital-on-a-carbon-atom-directed-along-the-POAV-and_fig6_235497518

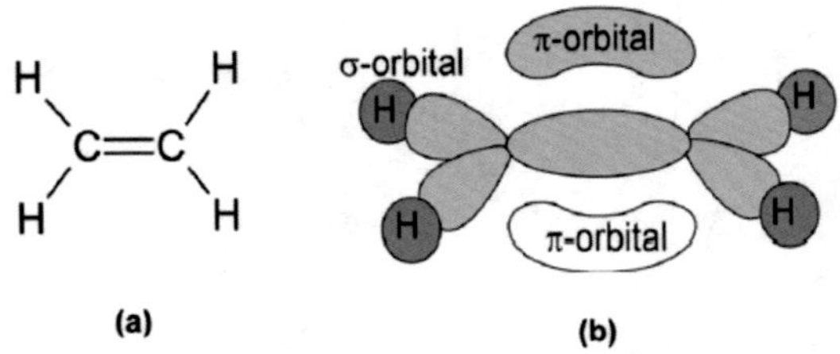

Figure 1.4. (a) Chemical structure of ethylene; and (b) schematic orbital diagram.

Source: https://www.researchgate.net/figure/Schematic-representation-of-the-most-important-molecular-orbitals-of-the-ethylene-radical_fig1_8441892.

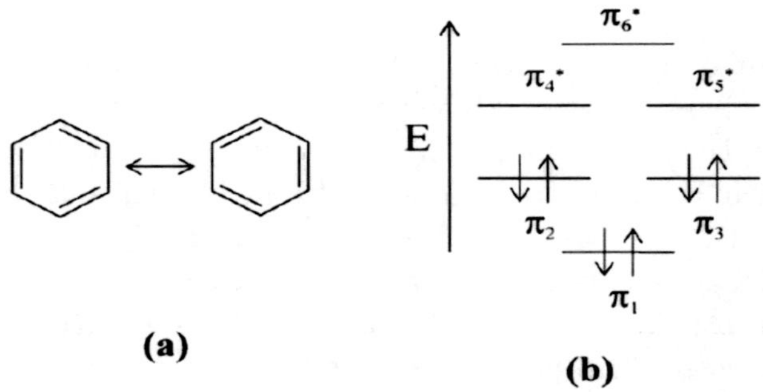

Figure 1.5. Two resonance structures and an energy map of the p-MOs make up the chemical composition of benzene.

Source: https://www.researchgate.net/figure/Calculated-molecular-orbitals-and-resonance-energies-for-the-s-aromaticity-of-dication-7_fig1_327917748.

The first conductive polymers to be found was trans-polyacetylene (Figure 1.6). The p-band, which is thought to be made up of an essentially limitless number of pz orbitals, would've been half-filled for just a single chain of neutrality polyacetylene, hypothetically leading to metallic behavior (Figure 1.6(a)). However, the findings of the experiments suggest that polyacetylene is a semiconductor with an energy gap of around 1.5 eV. As seen in Figure 1.6, the one-dimensional (1D) "metal" is unstable and has a propensity to deform (b). As a result, the polyacetylene longest chain consists of shorter and longer bonds that alternate (Chen and Chao, 2005; Horowitz, 2015). Trans-repeating polyacetylene's unit is changed from trans-$(CH)_n$ to trans-$(\text{-HC1/4CH-})_n$ via the so-called Peierls distortion. The heterodimer of the molecule creates an energy band gap, as seen in Figure 1.6(g). This group structure resembles that of normal inorganic semiconductors, which have a full-filled ring structure and an empty conduction band, with the p-band being completely occupied and the upper p*-band being vacant. Higher-dimensional p-electron materials are not affected by this Peierls instability, but 1D p-electron metals are not stable. Graphene and carbon nanotubes, for instance, may both be metallic or semi-metallic materials (Li et al., 2010).

1.3. SOLITONS, POLARONS, AND BIPOLARONS

The Su-Schrieffer-Heeger (SSH) model, which is founded on a relatively non-tight-binding model, may be used to characterize the electrical composition of such a polymer. It is assumed that the electron-phonon interaction between the p-electrons and the polymer framework is quite strong. Solitary waves, polaron, and bipolaron creation results in charge storage. The polymer chains' structural flaws are called a soliton. There are two resonance forms for polyacetylene, as shown in Figure 1.6(b) and (c). These two degenerate frequency configurations include a soliton as a domain border (Figure 1.6(d)); the domain wall may span multiple carbon atoms (Figure 1.6(e)). A soliton may either be neutral or charged. Because the solution is nonbonding, its energy is located between the p-p* gap. So, because system energy is independent of the soliton's location, this defective state is movable (Blochwitz et al., 2001; Kohlstedt et al., 2018). The solitons cannot be persistent for non-degenerate ground-state polymers, such as poly(phenylene vinylene) (PPV; Figure 1.7(a)) and polyparaphenylene (PPP,

Figure 1.7(b)), where the exchange of singles and double bonds results in energy shifts. Instead, such non-degenerate polymers need double defects. As a bonded state of charged solitary waves and a neutrality soliton, these double defects are known as polarons (Figure 1.7(c)). As a result, its energy levels combine the two conditions, which are divided into a bonding level with lower energy as well as an anti-bonding level with higher energy (Figure 1.7(d)). Positively or negatively polarons may occur when zero, one, or two electrons occupy one of several energy levels. A splitting pair of energy states between both the energy gap and two oppositely charged solitons with equal charges make up a bipolaron, which is a bound state of two of them. Additionally, structural relaxation is connected with solitons, polarons, and bipolarons, indicating significant electron-phonon connections of the charge storing in conjugated polymers. In those other worlds, lattice aberrations that may stabilize that system are present around the charges (Winkler and Houk, 2007; Yumusak et al., 2020).

1.4. EXCITONS

OSCs exhibit excitons as states of stimulated emission (Salzmann and Heimel, 2015). An electron/hole pair enclosed by the Coulomb force makes up an exciton, particle. An exciton has no net charge at all. However, an exciton may travel through organic molecules while carrying an excited state. Excitons may be classified as Frenkel, Wannier-Mott, or charge-transfer (CT) excitons (Figure 1.8). The bond length or distance here between electron/hole combinations is the major distinction between them. When the radius of the electron/hole pair is comparable with the size of a single molecule or the crystalline constant (aL), the exciton is of Frenkel type. The small radius also implies a strong coulombic interaction. In other words, it is very difficult to overcome the binding energy and dissociate the exciton into free charges. If the radius of the exciton is much larger than aL, it is a Wannier–Mott exciton. Therefore, the binding energy of this type of exciton is small. The intermediate case, in which the exciton extends over several molecular units, it is termed a CT exciton. (Lüssem et al., 2013; Zhao et al., 2020).

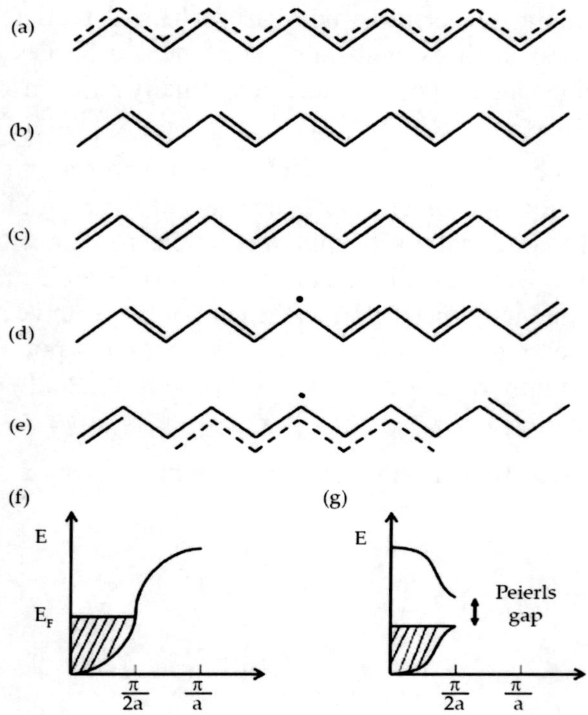

Figure 1.6. Polyacetylene is shown in figures (a–e) as follows: (a) an undimerized structure; (b, c) two resonance forms arising from Peierls effects; (d) a soliton in trans-polyacetylene; and (e) a soliton domain wall extension; (f, g) plots of the electronic band dispersion E(k) for dimerized and non-dimerized systems are shown in (f) and (g), respectively. E_f is the Fermi energy, and the dimerized system exhibits a gap known as the "Peierls gap."

Source: https://www.researchgate.net/figure/Two-conformations-of-polyacetylene-a-trans-PA-and-b-cis-PA_fig4_44230111.

Due to their significant binding energy, which results from the insulator loss constants of organic compounds and their very confined electronic clouds, the majority of excitons in OSCs are Frenkel-type. Excitons in OSCs typically have binding energies of about 0.5 eV (Biskup, 2021).

In organic optoelectronics, the production of excitons is a critical step. The combination of holes and electrons in OLEDs yields excitons, which later decompose to create EL. The excitons produced by absorption of energy in sustainably grown photovoltaic devices (like organic solar cells) move freely inside semiconductor materials because of their high compulsory energy, but they are topic to appropriate start charging detachment at the functionality among powerful carbon-based charged particle and organic electron-acceptors. Such excited-state detachment produces free charges, which, after being gathered by the electrodes, finally result in a photocurrent (Liu et al., 2020).

The orientations of spin of both the excitation energy aren't likely to alter easily when created in ordinary OSCs because of the high binding energy. According to the combinations of spin orientations in the excitons, there are two distinct sorts. The total spin rotational motion of a molecule in its initial state is represented in Figure 1.9(a) as the sum of two spin vectors, which is equal to zero due to the opposing spin orientations. Or to put it another way, the total rotation important quantity (S) is 0 (Lee et al., 2020).

$$S = |s1 + s2| = |(+1/2) + (-1/2)| = 0 \qquad (1)$$

where; the particular electron's angular momentum quantum numbers are s1 and s2.

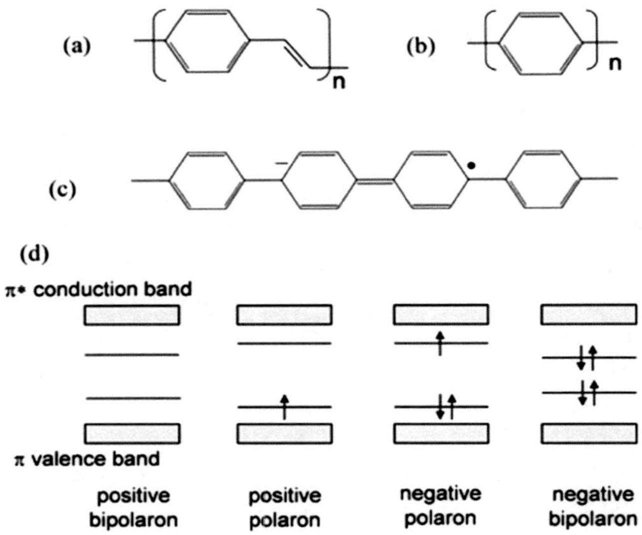

Figure 1.7. Chemical structure of (a) poly(phenylene vinylene) (PPV) and (b) polyparaphenylene (PPP).Poly(phenylene vinylene) (PPV) and poly para phenylene (PPP) chemical structures (PPP). (c) A negative polaron in PPP is shown schematically. (d) A band diagram showing different polarons.

Source: https://www.researchgate.net/publication/349003858_Conducting_polymers_a_comprehensive_review_on_recent_advances_in_synthesis_properties_and_applications.

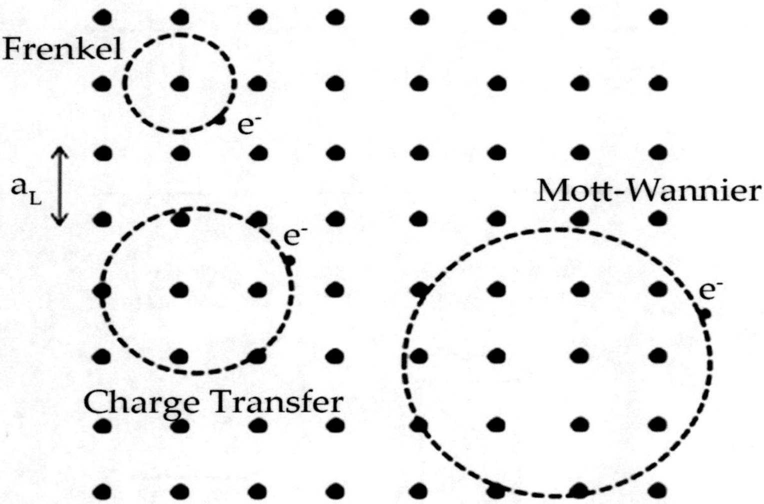

Figure 1.8. Three forms of excitons in solid-state resources are shown schematically; a_L stands for the lattice constant.

Source: https://www.researchgate.net/figure/a-Schematic-diagram-of-a-square-lattice-formed-by-coupled-exciton-polariton_fig1_339200132.

Because there is no net angular impetus in the molecule when one electron is stimulated to the p-orbital, the resultant exciton is known as a singlet (Figure 1.9(b)). The total rotation angular impetus is equal to 1 in contrast, as shown by the subsequent equation if the two spin vectors point in almost the same direction (Manousiadis et al., 2020).

$$S = |s1 + s2| = |(+1/2) + (+1/2)| = 1 \quad (2)$$

Notably, the total wouldn't change if s1 and s2 were both equal to–1/2. As a result, there is now a net rotational moment present in the excited singlet state (exciton). The exciton's energy might be divided into three distinct energy levels, providing there is a magnetic field. As a result, it is referred to as a triplet (Figure 1.9(c)).

Possible decay pathways of a prototypical organic molecule are shown in Figure 1.9(d), where S_0, S_1, and T_1 stand in for the ground state k_{nr} and k' are the average annual of non-radiative decay processes, while k_r and k' are the corresponding rates for the singlet stage and first triplet state (Karl, 2003; Gao et al., 2020). Inter-system crossover (ISC) is a process whose equilibrium constant is k_{isc}.

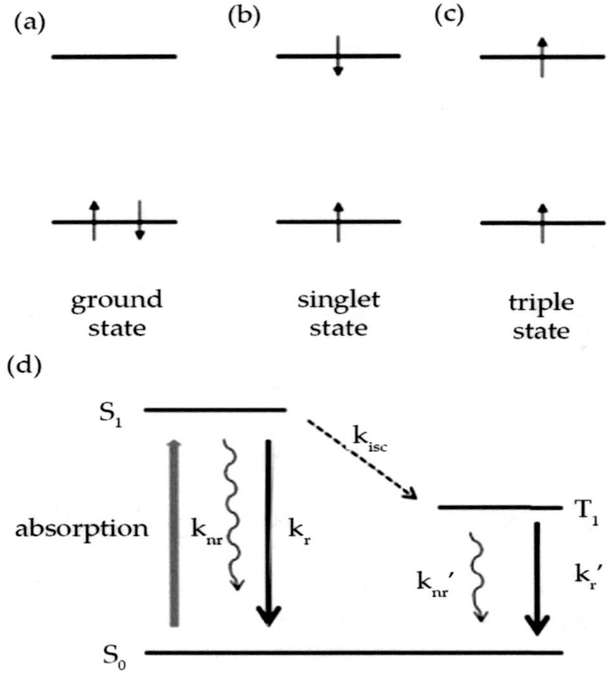

Figure 1.9. Electron spinning in an organic molecule's initial state, singlet exciton, and triplet exciton is shown schematically in (a), (b), and (c), respectively. (d) Processes of OSC excitation and potential decay.

Source: https://www.researchgate.net/figure/Schematic-representation-of-the-electronic-structure-for-the-singlet-ground-state-and-the_fig4_8347740.

As seen in Figure 1.9(d), the spin rotational motion is preserved, making it feasible for a singlet exciton (S_1) to relax to the ground state (S_0). Such a transition is permitted in quantum mechanics, and its radiation decay may result in the extremely effective emission known as fluorescence (k_r) (Karl, 2003). A triplet exciton cannot go from its excited state to its initial state, on the other hand. As a result, the radiation decay is exceedingly inefficient and, for the majority of organic compounds, is only detectable at very low temperatures. The term "phosphorescence" (k') refers to the discharge of triplet excitation energy (Roncali et al., 2007; Moral et al., 2017).

1.5. CONCEPT OF DOPING AND P- AND N-TYPE OSCS

OSCs may be categorized as p- and n-type semiconductors, much as inorganic semiconductors. Although the word "doping" is employed, the doping idea and procedure are quite different from those for inorganic compounds. For instance, phosphorus (P) or boron (B) atoms are injected as impurities during the ion implantation process to dope the semiconductor-based silicon. A few of the silicon atoms are replaced by these alien atoms, which creates new energy levels within the band gap. After the addition of *P* or *B* atoms, silicon changes into an n- or p-type semiconductor, depending (Leontie et al., 2005; Li et al., 2018).

The "doping" of OSCs, on the additional hand, can be talented through a diversity of methods, counting chemical nobbling, electrochemical doping, photo-doping, and custody inoculation from metal associates. As shown in the subsequent instances for polymers, chemical doping involves CT redox chemistry.

$$p-type:(\pi - polymer)_{\pm} + 3/2ny(I_2) \rightarrow [(\pi - polymer)^{+y}(I_3^-)\ ] \quad (3)$$

$$n-type:(\pi - polymer)_n + [Na^+(C_{10}H_8)^-]_y \rightarrow [(Na^+)_y(\pi - polymer)^{-y}]_n + (C_{10}H_8)^0 \quad (4)$$

p- and n-type doping are comparable to electrical and chemical reduction and oxidation, accordingly, from such a chemical perspective. In the OSCs, charged particles are produced via the doping process. When the amount of doping is increased, the dielectric strength rises and, at extremely high dopant concentration, may equal that of a metal. This chemical method's disadvantage is that it's difficult to manage the doping level, particularly for intermediate values. Thankfully, electrochemical doping can address this issue. To balance the charge mismatch diffuses from the dielectric fluid to or from the polymers at the same time as an electrode resources redox charges to the semiconductor. The exact doping level may be determined by controlling the electrode's voltage (Leontie et al., 2005; Scaccabarozzi et al., 2021). The following formula may be used to demonstrate electrochemical doping:

$$p-type:(\pi - polymer)_n + [Li^+(BF_4^-)]_{soln} \rightarrow [(\pi - polymer)^{+y}(BF_4^-)]_n + Li_{electrode} \quad (5)$$

$$n-type:(\pi - polymer)_n + Li_{electrode} \rightarrow [(Li^+)_y(\pi - polymer)^{-y}]_n \rightarrow [Li^+(BF_4^-)]_{soln} \quad (6)$$

The third technique, called photo-doping, involves generating a charged particle pair by absorbing light, which may cause the material to become regionally oxidized (p-type doping) or neighboring decreased (n-type doping). The following formula illustrates the special mechanism:

$$(\pi - polymer)_n + hv \rightarrow [(\pi - polymer)^{+y} + (\pi - polymer)^{-y}]_n \quad (7)$$

Rearrangements to the initial state after light absorption can either be convective or nonradiative. The excitation rate and recombination ratio compete for control over the y value in the formula. In the last situation, metallic contacts pump charge into the OSCs. The following equation indicates how holes and electrons are injected into the p- and p*-bands, correspondingly:

$$p-type:(\pi-polymer)_n-y(e^-)\rightarrow[(\pi-polymer)^{+y}]_n \qquad (8)$$

$$n-type:(\pi-polymer)_n+y(e^-)\rightarrow[(\pi-polymer)^{-y}]_n \qquad (9)$$

Eqn. (8) depicts the introduction of such a hole, equivalent to the oxidation process, through into the filled p-band (elimination of an electron) (Köhler and Bässler, 2011).

Eqn. (9) depicts the equivalent of a reduction reaction—the insertion of the electron into such an empty p*-band. The lack of counter ions to balance the charges is the primary differentiator between electrode potential doping and other doping procedures (such as chemical doping). The phases are thus highly unstable, and the doping procedure is only transitory. Contrarily, unless the transporters are eliminated by an "undoing" procedure, chemical or electrochemical doping is irreversible. The photoconductivity that results from photo-doping is momentary and constrained by mixing and decaying mechanisms (Liao and Yan, 2013; Zheng and Huo, 2021).

Although OSCs may be "doping" as previously said in order of becoming p- or n-type semiconductors, several studies characterize OSCs as p- or n-type by determining if they are effective conductors of electron-hole pairs. For example, due to its extremely rapid hole mobility, pentacene is often considered a superb p-type semiconductor (Figure 1.2(d)). In OLEDs, TPD (Figure 1.2(d)) is often used as a material that transports holes. Because TPD successfully carries holes but also because holes are easily inserted from the metallic conductors, it is sometimes referred to as a p-type OSC. It should be noted that such OSCs are not purposefully doped, and before processing for device construction, they are often highly cleaned to get rid of contaminants. Therefore, unlike inorganic semiconductors, we cannot assume that additional energy states are usually present in the HOMOs. Determining whether OSCs operate as either electron donors (p-type) or electron acceptors (n-type) throughout transient conditions is another method of characterizing them. For instance, fullerene analogs (Figure 1.2(e)) are referred to be n-type OSCs because they carry photogenerated photovoltaic devices and take electrons on photo-excitation. The fact that several conjugated polymers combined with fullerene compounds act as p-type OSCs during photoexcitation in solar panels leads to the classification of these materials (Stingelin, 2012; Scheunemann et al., 2020).

1.6. DEVICE APPLICATIONS

This unit will review several of the most significant organic electronic components, such as organically light-emitting diodes (OLEDs), organic thin-film transistors (OTFTs), and electrophotographic sensors (OLETs). These are frequently used in workplaces with copiers and laser printers. Organic photoconductors were the first innovation to incorporate OSCs as a huge element in commercially accessible goods (Figure 1.10). With such an electrophotographic system, a collection of optical components transfers the images of a paper to a photoconductive plate (photoreceptor), which has already been charged by electrostatic discharge (Brown et al., 1997; Lin and Corminboeuf, 2020). Due to the photoconductive plate's strong photoconductivity, just a portion of it discharges when exposed to light, leaving a duplicated image of both the actual document on its surface. The electrostatic interactions charged toner particles are applied over the partly charged surface in the following phase of the process; a few of the toner particles are attached to a picture by the electromagnetic field. The toner-particle-based picture is subsequently transferred to a piece of paper by an electrostatic field. Lastly, pressure and heating are used to fix the picture. The photoconductive plates are prepared for the subsequent copy procedure after being cleaned of any leftover toner grains and erased by exposing that to light (Lee et al., 2010).

The OLED's most basic device construction is shown in Figure 1.11(a). A hole-transport membrane, as well as an electron-transport layer (ETL), are its usual two organic levels (HTL). Finding an OSC with balanced input and transit among electron-hole pairs is not necessary with this arrangement. The bilayer system makes it easier for holes to be injected from such indium tin oxide (ITO) electrodes and electrons to be injected from a metal or alloy (MgAg) electrode than a single crystal (such as anthracene) does. The thickness of the HTLs and ETLs may be adjusted to further fine-tune the charge balance. After the infused electron-hole pairs recombined, the excitons that were created relaxed to the dipole moments, causing photon discharge. Strong green emissions are seen through the ITO surface of such a device due to the transparency of the ITO anode (Van Der Holst et al., 2009; Armin et al., 2017).

Only extremely weak van der Waals interactions be between both the molecules in solid-state organic materials. However, several experimental findings have shown that p-electrons may be efficiently transported through OSCs. The charge transport characteristics of OSCs may be investigated using organic field effect transistors (OFETs). Contemporary electrical circuits rely heavily on transistors, which may be either signal amplification or on/off switches. The "field effect" describes the phenomenon where a semiconductor's conductance is altered by introducing an electromagnetic current perpendicular to its surface. Since the majority of OFETs are OTFTs, the dynamic (semiconducting) layer must be almost two-dimensional. The three typical device architectures of OTFTs are shown in Figure 1.11(b) through (d), and the operating mechanism is shown in Figure 1.11(e). The shows that a number at the dielectric-OSC move down a channel boundary.

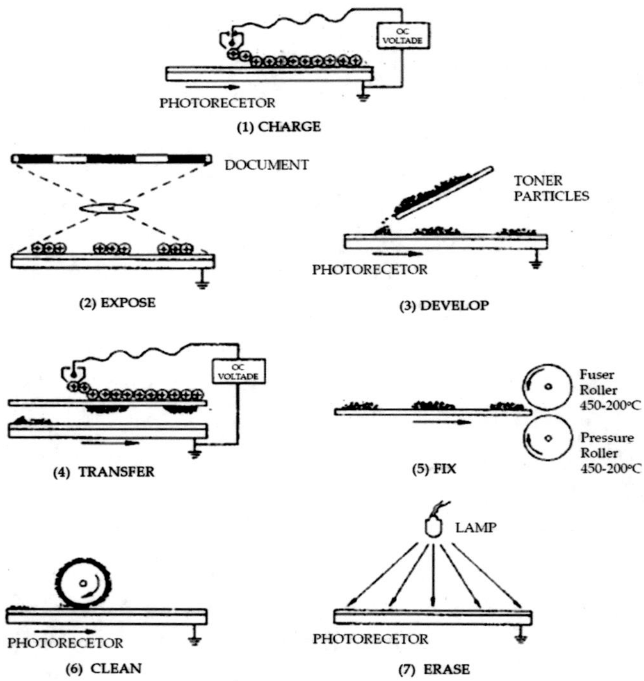

Figure 1.10. Working principle of an electrophotographic device: (1) charge; (2) depiction; (3) grow; (4) transmission; (5) fix; (6) clean; and (7) erase.

Source: https://aip.scitation.org/doi/10.1063/1.4918556.

The source/drain electrodes give the OSCs contacts and may inject or gather the charges within the channel, respectively. A third gate conductor that is spaced apart by a coating of the barrier layer may change the conductivity of the channel (Anthony et al., 2008; Scholdt et al., 2010). For instance, a negative voltage supplied to the gate (VG) of a p-channel OTFT causes holes to form in the channel due to field effects. The field-induced holes are driven to migrate from the sources to the drainage electrodes by the addition of a negative bias between both the source and drain terminals (VD). The device behaves as a resistor, with the drain current (ID) cumulative linearly concerning the increasing value of VD. This operating zone is known as the linear range because of the link between ID and VD (Figure 1.11(f)). The magnitude of both the field that induces the charges is partly canceled at the drain even if a significant amount of VD is provided. The proportion of generated charges decreases to nearly zero whenever the functional voltage falls below the threshold voltage (Vt), resulting in the formation of an increased confrontation rule within the channel towards the channel end. The ID value is now beginning to saturation, as well as the channel is now pinched at one side. The operating zone in question is the saturated region, and so this phenomenon is referred to as "pinch-off" (Baumeier et al., 2012; Wang et al., 2020) (Figure 1.11(f)).

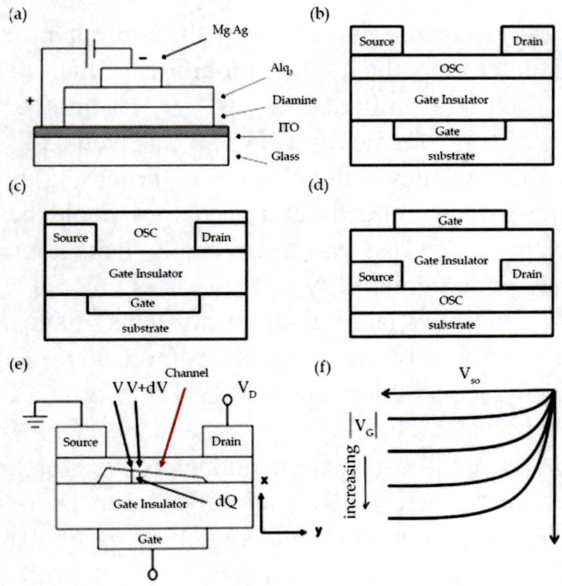

Figure 1.11. The coatings of Alq3 and diamine are the electron-transport layer (ETL) and hole-transport layer (HTL), respectively, in (a), which shows the device structure of a typical organic light-emitting diode (OLED). Organic thin-

film transistors (OTFTs) often have the following structures; (b) top-contact, bottom-gate; (c) bottom-contact, bottom-gate; and (d) bottom-contact, top-gate. (e) An illustration of how a typical OTFT operates. (f) A typical OTFT's output characteristics (ID-VD) [ITO: Indium tin oxide].

Source: https://www.researchgate.net/figure/OLED-performance-a-Device-structure-and-ionization-potentials-and-electron-affinities_fig4_334476413.

Importantly, the supplied gate voltage affects the saturated current's size. The electrical behavior of n-channel OTFTs is similar, however, the existing curves are situated from the first rather than the third quadrant. Therefore, electrons are the charge carriers (Mikhnenko et al., 2015).

The drain flow there in the linear area is defined as follows:

$$I_D = C_{ox\mu}(W / L)[(V_G - V_{th})V_D - (1/2)V_D{}^2] \quad (10)$$

where; C_{ox}, m is the OSC's movement; W and L are the channel's length and breadth, and C is the capacitor per unit of area of a dielectric substrate. The quantity of ID may be stated as follows in the saturation mode:

$$I_D = C_{ox\mu}(W / 2L)[(V_G - V_{th})^2 \quad (11)$$

As a result, by fitting the curves using the aforementioned formulas, it is possible to determine the OSCs' mobilities either in the linear or saturating areas. The on/off ratio denoted as I_{on}/I_{off}, is another crucial factor for evaluating the effectiveness of an OTFT. I_{on}, as well as I_{off}, are the drain currents, respectfully, whenever the channel is turned on and off. Charges may leak past the gate or indeed certain charges could be stored inside semiconductors, which can result in a detectable leakage current. Modern OTFTs may easily achieve mobility of up to 1 cm^2 V^{-1} s^{-1}. For solution-processed OTFTs, mobilities larger than 10 cm^2 V^{-1} s^{-1} have been observed. For instance, a highly aligned metastable molecule of 2,7-dioctyl[1]benzothieno [3,2-b][1]benzothiophene (C8-BTBT), formed from a mixture of C8-BTBT and polystyrene, has shown a velocity of 43 cm^2 V^{-1} s^{-1} (on average 25 cm^2 V^{-1} s^{-1}), indicating tremendous promise for future electrical applications (Neupane et al., 2019). Field-effect transistors are necessary for controlling the luminosity levels and switching particular OLED pixels whenever OLEDs are used as components of active-matrix display screens since they are generally two-terminal transistors. In comparison, whenever OLETs are used for comparable applications, no transistors are required since a single OLET can carry out both light-generating and electrically switching capabilities. In a unipolar OTFT, the conductive channel is always injected with one kind of charge carrier, either an electron or a hole. But for OLETs to emit light, excitons must develop between two carriers. OLETs that are unipolar were also observed, however, ambipolar devices are preferred. Unipolar OLETs emit light when an electrode is nearby, and since the excitons are easily quenched by the metals, this might reduce the luminescent efficiency (Lee et al., 2010; Blakesley et al., 2014).

Figure 1.12(a) exhibits a standard OLET with a bottom connection geometry; it resembles an ordinary OTFT in construction. To put it another way, additional structures, such as the two additional geometries in Figure 1.11, are always an option. The OSC must be emissive, and the source/drain electrodes are injected with holes and electrons, respectively. Hole or electron injection occurs from the drain and source connections, correspondingly, in response to the application of an acceptable gate bias (VG). The two opposing charges combine to create excitons within the channel, which then emit light by radiative decay (D'Andrade et al., 2005; Li et al., 2007).

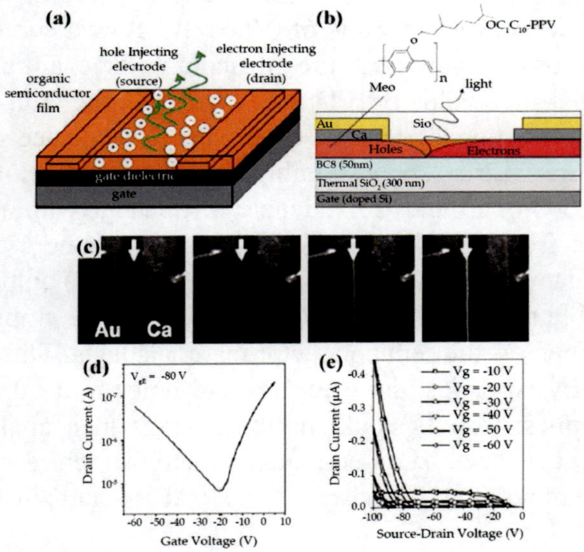

Figure 1.12. Diagram representation of a characteristic carbon-based light-emitting transistor (OLET) device structure and its working principle.

Source: https://www.researchgate.net/figure/a-Schematic-of-the-device-structure-and-working-principle-of-a-typical-planar-OLET-with_fig1_351784959.

In reality, it may be quite challenging to get an ambipolar technology to work at its optimal level. The sources and drain must first be able to effectively introduce both holes as well as electrons. Additionally, the OSC need to have equal hole and electrode high mobility. The first ambipolar OLETs using poly(2-methoxy-5-(3,7-dimethyloctoxy)-p-phenylene-vinylene) (MDMO-PPV), linked polymers with high photoluminescence (PL) effectiveness, were reported by Zaumseil et al. in 2006 (Figure 1.12(b)). To stop electrons from being trapped at the dielectric-semiconductor interface, they placed a coating of crosslinked benzocyclobutene derivative (BCB) just on the dielectric layer. Another of the secrets to enhancing electron (n-channel) transit on oxide insulating surfaces is this strategy. Then, for an efficient hole or electron insertion, correspondingly, Au and Ca electrodes were formed. Only using Au electrodes both for connections would result in such a side with a high electrical injection barrier. Similar to this, if Ca electrodes had been used exclusively, one of the connections would have had a high hole injection barrier. The motion of the light emitting zone between both the two electrodes as well as the light emission from the OLET are shown in Figure 1.12(c). In reality, due to restricted electron injection, there were no light emissions whenever the value of |VD| fell below a certain threshold. That once value of |VD| surpassed the threshold voltage, a thin line of light became visible. As the value of |VD| was increased, the light emitting line traveled in the direction of the Au electrodes. The device's transferring properties (Figure 1.12(d)) showed ambipolar behavior; the lowest point, at such a value of VG of around–20 V, denotes the change within the current's dominant mode from one dominated by electrons to one by holes in the OLET. Ambipolar currents are also seen in Figure 1.12(e) output properties. It is also possible to create high-performance OLETs by employing mixed p- and n-type OSCs as the semiconductor materials layer. This method may be carried out by co-depositing electrons and holes in a bilayer structure conveying supplies. OLETs could perform better than analogous LEDs when utilizing a trilayer semi-conducting heterostructure, according to much more recent research (Rand et al., 2007; Abate et al., 2013).

1.7. SUMMARY

In several ways, the characteristics of OSCs vary significantly from those of semiconductor materials. For instance, the charge transport in OSCs is carried out by p-electrons, therefore the idea of doping is significantly different. Despite this, OSCs are very desirable for application in electrical devices since they are created from cheap precursor material and offer a variety of advantageous properties, such as mobility and a way to solve. Numerous opportunities for different applications are provided by these special qualities. For instance, huge devices are exceedingly challenging to execute, and places with high LEDs are often made by careful crystal formation. OLEDs, on the other hand, are easily pixelated and may be utilized for flat-panel screens. However, it is important to note that organic electrical devices aren't created to outperform those having inorganic equivalents. Instead, we think there would be a lot of new uses where inorganic semiconductors won't be practical, thus more research into OEs is necessary for the future (Quinn et al., 2017; Alessandri et al., 2020).

REFERENCES

1. Abate, A., Leijtens, T., Pathak, S., Teuscher, J., Avolio, R., Errico, M. E., & Snaith, H. J., (2013). Lithium salts as "redox active" p-type dopants for organic semiconductors and their impact in solid-state dye-sensitized solar cells. *Physical Chemistry Chemical Physics, 15*(7), 2572–2579.
2. Abate, A., Staff, D. R., Hollman, D. J., Snaith, H. J., & Walker, A. B., (2014). Influence of ionizing dopants on charge transport in organic semiconductors. *Physical Chemistry Chemical Physics, 16*(3), 1132–1138.
3. Alessandri, R., Sami, S., Barnoud, J., De Vries, A. H., Marrink, S. J., & Havenith, R. W., (2020). Resolving donor–acceptor interfaces and charge carrier energy levels of organic semiconductors with polar side chains. *Advanced Functional Materials, 30*(46), 2004799.
4. Allard, S., Forster, M., Souharce, B., Thiem, H., & Scherf, U., (2008). Organic semiconductors for solution-processable field-effect transistors (OFETs). *Angewandte Chemie International Edition, 47*(22), 4070–4098.
5. Ando, S., Nishida, J. I., Fujiwara, E., Tada, H., Inoue, Y., Tokito, S., & Yamashita, Y., (2005). Novel p-and n-type organic semiconductors with an anthracene unit. *Chemistry of Materials, 17*(6), 1261–1264.
6. Anthony, J. E., Heeney, M., & Ong, B. S., (2008). Synthetic aspects of organic semiconductors. *MRS Bulletin, 33*(7), 698–705.
7. Armin, A., Stoltzfus, D. M., Donaghey, J. E., Clulow, A. J., Nagiri, R. C. R., Burn, P. L., & Meredith, P., (2017). Engineering dielectric constants in organic semiconductors. *Journal of Materials Chemistry C, 5*(15), 3736–3747.
8. Bässler, H., & Köhler, A., (2011). Charge transport in organic semiconductors. *Unimolecular and Supramolecular Electronics I,* 1–65.
9. Baumeier, B., May, F., Lennartz, C., & Andrienko, D., (2012). Challenges for in silico design of organic semiconductors. *Journal of Materials Chemistry, 22*(22), 10971–10976.
10. Biskup, T., (2021). Doping of organic semiconductors: Insights from EPR spectroscopy. *Applied Physics Letters, 119*(1), 010503.
11. Blakesley, J. C., Castro, F. A., Kylberg, W., Dibb, G. F., Arantes, C., Valaski, R., & Kim, J. S., (2014). Towards reliable charge-mobility

benchmark measurements for organic semiconductors. *Organic Electronics, 15*(6), 1263–1272.

12. Blochwitz, J., Fritz, T., Pfeiffer, M., Leo, K., Alloway, D. M., Lee, P. A., & Armstrong, N. R., (2001). Interface electronic structure of organic semiconductors with controlled doping levels. *Organic Electronics, 2*(2), 97–104.
13. Brédas, J. L., & Marder, S. R., (2016). *The WSPC Reference on Organic Electronics: Organic Semiconductors* (Vol. 1). Singapore: World Scientific.
14. Bronstein, H., Nielsen, C. B., Schroeder, B. C., & McCulloch, I., (2020). The role of chemical design in the performance of organic semiconductors. *Nature Reviews Chemistry, 4*(2), 66–77.
15. Brown, A. R., Jarrett, C. P., De Leeuw, D. M., & Matters, M., (1997). Field-effect transistors made from solution-processed organic semiconductors. *Synthetic Metals, 88*(1), 37–55.
16. Brütting, W., (2005). Introduction to the physics of organic semiconductors. *Physics of Organic Semiconductors*, 1–14.
17. Chen, H. Y., & Chao, I., (2005). Effect of perfluorination on the charge-transport properties of organic semiconductors: Density functional theory study of perfluorinated pentacene and sexithiophene. *Chemical Physics Letters, 401*(4–6), 539–545.
18. D'Andrade, B. W., Datta, S., Forrest, S. R., Djurovich, P., Polikarpov, E., & Thompson, M. E., (2005). Relationship between the ionization and oxidation potentials of molecular organic semiconductors. *Organic Electronics, 6*(1), 11–20.
19. Dong, H., Wang, C., & Hu, W., (2010). High performance organic semiconductors for field-effect transistors. *Chemical Communications, 46*(29), 5211–5222.
20. Forrest, S. R., (2000). Active optoelectronics using thin-film organic semiconductors. *IEEE Journal of Selected Topics in Quantum Electronics, 6*(6), 1072–1083.
21. Gao, Y., Wu, Y., Liu, Y., Lu, M., Yang, L., Wang, Y., & Dai, Q., (2020). Interface and grain boundary passivation for efficient and stable perovskite solar cells: The effect of terminal groups in hydrophobic fused benzothiadiazole-based organic semiconductors. *Nanoscale Horizons, 5*(12), 1574–1585.

22. He, Y., Xu, W., Murtaza, I., Zhang, D., He, C., Zhu, Y., & Meng, H., (2016). Molecular phase engineering of organic semiconductors based on a [1] benzothieno [3, 2-b][1] benzothiophene core. *RSC Advances, 6*(97), 95149–95155.
23. Henson, Z. B., Müllen, K., & Bazan, G. C., (2012). Design strategies for organic semiconductors beyond the molecular formula. *Nature Chemistry, 4*(9), 699–704.
24. Hofmann, A. I., Kroon, R., & Müller, C., (2019). Doping and processing of organic semiconductors for plastic thermoelectrics. In: *Handbook of Organic Materials for Electronic and Photonic Devices* (pp. 429–449). Woodhead Publishing.
25. Horowitz, G., (2015). Validity of the concept of band edge in organic semiconductors. *Journal of Applied Physics, 118*(11), 115502.
26. Hu, P., He, X., & Jiang, H., (2021). Greater than 10 cm2 V− 1 s− 1: A breakthrough of organic semiconductors for field-effect transistors. *InfoMat, 3*(6), 613–630.
27. Jacob, M. V., (2014). Organic semiconductors: Past, present and future. *Electronics, 3*(4), 594–597.
28. Jiang, W., Li, Y., & Wang, Z., (2013). Heteroarenes as high performance organic semiconductors. *Chemical Society Reviews, 42*(14), 6113–6127.
29. Karl, N., (2003). Charge carrier transport in organic semiconductors. *Synthetic Metals, 133*, 649–657.
30. Knupfer, M., (2003). Exciton binding energies in organic semiconductors. *Applied Physics A, 77*(5), 623–626.
31. Köhler, A., & Bässler, H., (2009). Triplet states in organic semiconductors. *Materials Science and Engineering: R: Reports, 66*(4–6), 71–109.
32. Köhler, A., & Bässler, H., (2011). What controls triplet exciton transfer in organic semiconductors?. *Journal of Materials Chemistry, 21*(12), 4003–4011.
33. Kohlstedt, K. L., Jackson, N. E., Savoie, B. A., & Ratner, M. A., (2018). Introduction to organic semiconductors using accessible undergraduate chemistry concepts. *Journal of Chemical Education, 95*(9), 1500–1511.

34. Krause, S., Casu, M. B., Schöll, A., & Umbach, E., (2008). Determination of transport levels of organic semiconductors by UPS and IPS. *New Journal of Physics, 10*(8), 085001.

35. Lee, J. H., Leem, D. S., & Kim, J. J., (2010). Effect of host organic semiconductors on electrical doping. *Organic Electronics, 11*(3), 486–489.

36. Lee, J., Park, S. A., Ryu, S. U., Chung, D., Park, T., & Son, S. Y., (2020). Green-solvent-processable organic semiconductors and future directions for advanced organic electronics. *Journal of Materials Chemistry A, 8*(41), 21455–21473.

37. Lee, W. H., Cho, J. H., & Cho, K., (2010). Control of mesoscale and nanoscale ordering of organic semiconductors at the gate dielectric/semiconductor interface for organic transistors. *Journal of Materials Chemistry, 20*(13), 2549–2561.

38. Leontie, L., Druta, I., Danac, R., Prelipceanu, M., & Rusu, G. I., (2005). Electrical properties of some new high resistivity organic semiconductors in thin films. *Progress in Organic Coatings, 54*(3), 175–181.

39. Li, J. L., Cao, J. J., Duan, L. L., & Zhang, H. L., (2018). Evolution of isoindigo-based electron-deficient units for organic electronics: From natural dyes to organic semiconductors. *Asian Journal of Organic Chemistry, 7*(11), 2147–2160.

40. Li, L., Meller, G., & Kosina, H., (2007). Carrier concentration dependence of the mobility in organic semiconductors. *Synthetic Metals, 157*(4, 5), 243–246.

41. Li, R., Hu, W., Liu, Y., & Zhu, D., (2010). Micro-and nanocrystals of organic semiconductors. *Accounts of Chemical Research, 43*(4), 529–540.

42. Liao, C., & Yan, F., (2013). Organic semiconductors in organic thin-film transistor-based chemical and biological sensors. *Polymer Reviews, 53*(3), 352–406.

43. Lin, K. H., & Corminboeuf, C., (2020). FB-REDA: Fragment-based decomposition analysis of the reorganization energy for organic semiconductors. *Physical Chemistry Chemical Physics, 22*(21), 11881–11890.

44. Lin, Y., Fan, H., Li, Y., & Zhan, X., (2012). Thiazole-based organic semiconductors for organic electronics. *Advanced Materials, 24*(23), 3087–3106.
45. Liu, Q., Bottle, S. E., & Sonar, P., (2020). Developments of diketopyrrolopyrrole-dye-based organic semiconductors for a wide range of applications in electronics. *Advanced Materials, 32*(4), 1903882.
46. Liu, Z., Zhang, G., Cai, Z., Chen, X., Luo, H., Li, Y., & Zhang, D., (2014). New organic semiconductors with imide/amide-containing molecular systems. *Advanced Materials, 26*(40), 6965–6977.
47. Lüssem, B., Riede, M., & Leo, K., (2013). Doping of organic semiconductors. *Physica Status Solidi (a), 210*(1), 9–43.
48. Ma, Z., Geng, H., Wang, D., & Shuai, Z., (2016). Influence of alkyl side-chain length on the carrier mobility in organic semiconductors: Herringbone vs. pi–pi stacking. *Journal of Materials Chemistry C, 4*(20), 4546–4555.
49. Manousiadis, P. P., Yoshida, K., Turnbull, G. A., & Samuel, I. D., (2020). Organic semiconductors for visible light communications. *Philosophical Transactions of the Royal Society A, 378*(2169), 20190186.
50. Meng, Q., Gao, J., Li, R., Jiang, L., Wang, C., Zhao, H., & Hu, W., (2009). New type of organic semiconductors for field-effect transistors with carbon-carbon triple bonds. *Journal of Materials Chemistry, 19*(10), 1477–1482.
51. Mikhnenko, O. V., Blom, P. W., & Nguyen, T. Q., (2015). Exciton diffusion in organic semiconductors. *Energy & Environmental Science, 8*(7), 1867–1888.
52. Moral, M., Pérez-Jiménez, A. J., & Sancho-Garcia, J. C., (2017). Understanding and controlling chemical modifications of rubicene for their envisioned use as molecular organic semiconductors. *The Journal of Physical Chemistry C, 121*(6), 3171–3181.
53. Myers, J. D., & Xue, J., (2012). Organic semiconductors and their applications in photovoltaic devices. *Polymer Reviews, 52*(1), 1–37.
54. Nakano, M., Osaka, I., Hashizume, D., & Takimiya, K., (2015). α-modified naphthodithiophene diimides- molecular design strategy for air-stable n-channel organic semiconductors. *Chemistry of Materials, 27*(18), 6418–6425.

55. Neupane, G. P., Ma, W., Yildirim, T., Tang, Y., Zhang, L., & Lu, Y., (2019). 2D organic semiconductors, the future of green nanotechnology. *Nano Materials Science, 1*(4), 246–259.

56. Newman, C. R., Frisbie, C. D., Da Silva, F. D. A., Brédas, J. L., Ewbank, P. C., & Mann, K. R., (2004). Introduction to organic thin film transistors and design of n-channel organic semiconductors. *Chemistry of Materials, 16*(23), 4436–4451.

57. Nitti, A., Forti, G., Bianchi, G., Botta, C., Tinti, F., Gazzano, M., & Pasini, D., (2021). Anthradithiophene-based organic semiconductors through regiodirected double annulations. *Journal of Materials Chemistry C, 9*(29), 9302–9308.

58. Okamoto, T., Suzuki, T., Tanaka, H., Hashizume, D., & Matsuo, Y., (2012). Tetracene dicarboxylic imide and its disulfide: Synthesis of ambipolar organic semiconductors for organic photovoltaic cells. *Chemistry–An Asian Journal, 7*(1), 105–111.

59. Pecile, C., Palnelli, A., & Girlando, A., (1989). Studies of organic semiconductors for 40 years—V. *Molecular Crystals and Liquid Crystals, 171*(1), 69–87.

60. Qi, T., Liu, Y., Qiu, W., Zhang, H., Gao, X., Liu, Y., & Zhu, D., (2008). Synthesis and properties of fluorene or carbazole-based and dicyanovinyl-capped n-type organic semiconductors. *Journal of Materials Chemistry, 18*(10), 1131–1138.

61. Quinn, J. T., Zhu, J., Li, X., Wang, J., & Li, Y., (2017). Recent progress in the development of n-type organic semiconductors for organic field effect transistors. *Journal of Materials Chemistry C, 5*(34), 8654–8681.

62. Rand, B. P., Genoe, J., Heremans, P., & Poortmans, J., (2007). Solar cells utilizing small molecular weight organic semiconductors. *Progress in Photovoltaics: Research and Applications, 15*(8), 659–676.

63. Roncali, J., Leriche, P., & Cravino, A., (2007). From one-to three-dimensional organic semiconductors: In search of the organic silicon?. *Advanced Materials, 19*(16), 2045–2060.

64. Salzmann, I., & Heimel, G., (2015). Toward a comprehensive understanding of molecular doping organic semiconductors. *Journal of Electron Spectroscopy and Related Phenomena, 204*, 208–222.

65. Scaccabarozzi, A. D., Basu, A., Aniés, F., Liu, J., Zapata-Arteaga, O., Warren, R., & Anthopoulos, T. D., (2021). Doping approaches for organic semiconductors. *Chemical Reviews, 122*(4), 4420–4492.

66. Scheunemann, D., Vijayakumar, V., Zeng, H., Durand, P., Leclerc, N., Brinkmann, M., & Kemerink, M., (2020). Rubbing and drawing: Generic ways to improve the thermoelectric power factor of organic semiconductors?. *Advanced Electronic Materials, 6*(8), 2000218.
67. Scholdt, M., Do, H., Lang, J., Gall, A., Colsmann, A., Lemmer, U., & Boettner, H., (2010). Organic semiconductors for thermoelectric applications. *Journal of Electronic Materials, 39*(9), 1589–1592.
68. Schwarze, M., Tress, W., Beyer, B., Gao, F., Scholz, R., Poelking, C., & Leo, K., (2016). Band structure engineering in organic semiconductors. *Science, 352*(6292), 1446–1449.
69. Sergeyev, S., Pisula, W., & Geerts, Y. H., (2007). Discotic liquid crystals: A new generation of organic semiconductors. *Chemical Society Reviews, 36*(12), 1902–1929.
70. Shen, Y., Hosseini, A. R., Wong, M. H., & Malliaras, G. G., (2004). How to make ohmic contacts to organic semiconductors. *ChemPhysChem, 5*(1), 16–25.
71. Sokolov, A. N., Cao, Y., Johnson, O. B., & Bao, Z., (2012). Mechanistic considerations of bending-strain effects within organic semiconductors on polymer dielectrics. *Advanced Functional Materials, 22*(1), 175–183.
72. Stingelin, N., (2012). On the phase behavior of organic semiconductors. *Polymer International, 61*(6), 866–873.
73. Sun, M., Xu, R., Xie, L., Wei, Y., & Huang, W., (2015). Toward eco-friendly green organic semiconductors: Recent advances in spiro [fluorene-9, 9′-xanthene](SFX)-based optoelectronic materials and devices. *Chinese Journal of Chemistry, 33*(8), 815–827.
74. Takimiya, K., Shinamura, S., Osaka, I., & Miyazaki, E., (2011). Thienoacene-based organic semiconductors. *Advanced Materials, 23*(38), 4347–4370.
75. Takimiya, K., Yamamoto, T., Ebata, H., & Izawa, T., (2007). Design strategy for air-stable organic semiconductors applicable to high-performance field-effect transistors. *Science and Technology of Advanced Materials, 8*(4), 273.
76. Torabi, S., Jahani, F., Van, S. I., Kanimozhi, C., Patil, S., Havenith, R. W., & Koster, L. J. A., (2015). Strategy for enhancing the dielectric constant of organic semiconductors without sacrificing charge carrier

mobility and solubility. *Advanced Functional Materials, 25*(1), 150–157.

77. Turkoglu, G., Cinar, M. E., & Ozturk, T., (2019). Thiophene-based organic semiconductors. *Sulfur Chemistry*, 79–123.
78. Van, D. H. J. J. M., Uijttewaal, M. A., Ramachandhran, B., Coehoorn, R., Bobbert, P. A., De Wijs, G. A., & De Groot, R. A., (2009). Modeling and analysis of the three-dimensional current density in sandwich-type single-carrier devices of disordered organic semiconductors. *Physical Review B, 79*(8), 085203.
79. Wang, Y., Gong, Q., & Miao, Q., (2020). Structured and functionalized organic semiconductors for chemical and biological sensors based on organic field effect transistors. *Materials Chemistry Frontiers, 4*(12), 3505–3520.
80. Winkler, M., & Houk, K. N., (2007). Nitrogen-rich oligoacenes: Candidates for n-channel organic semiconductors. *Journal of the American Chemical Society, 129*(6), 1805–1815.
81. Yamada, H., Okujima, T., & Ono, N., (2008). Organic semiconductors based on small molecules with thermally or photochemically removable groups. *Chemical Communications*, (26), 2957–2974.
82. Yamashita, Y., (2009). Organic semiconductors for organic field-effect transistors. *Science and Technology of Advanced Materials, 10*(2), 024313.
83. Yumusak, C., Sariciftci, N. S., & Irimia-Vladu, M., (2020). Purity of organic semiconductors as a key factor for the performance of organic electronic devices. *Materials Chemistry Frontiers, 4*(12), 3678–3689.
84. Zhao, W., Ding, J., Zou, Y., Di, C. A., & Zhu, D., (2020). Chemical doping of organic semiconductors for thermoelectric applications. *Chemical Society Reviews, 49*(20), 7210–7228.
85. Zheng, B., & Huo, L., (2021). Recent advances of dithienobenzodithiophene-based organic semiconductors for organic electronics. *Science China Chemistry, 64*(3), 358–384.

CHAPTER 2

STRUCTURE AND PROPERTIES OF ORGANIC SEMICONDUCTORS

CONTENTS

2.1. INTRODUCTION

There are two types of organic semiconductors (OSC): polymers and specific molecular components. There is a significant chemical distinction between such two material types, which are also mirrored in technical capabilities. Polymeric macromolecules are formed by the repeating of a basic unit, the mono, and therefore are dissolved in organic solvents, allowing them to be handled in a fluid state (Bässler, 1993; Blom, 2020). The latter group of components sometimes referred to as molecular components, consists of tiny molecules and could be split into two subsets: dyes, which are insoluble in organic liquids, and dyes, which have been dissolvable. Based on their chemical characteristics, polymers may be solution treated, such as being twisted from a mixture of suitable organic liquids, but small molecule substances should be heated vaporized and could be liquid produced in select instances. In this book, thermally vaporized molecular substances have been explored, but this chapter will provide a detailed introduction to any types of OSC, starting with a discussion of their forms and characteristics, accompanied by an explanation of their fundamentals. To consider organic solar cells beginning with inorganic foundations is a standard procedure, but it is frequently confusing: ideas such as n- and p-type semiconductors, the energizing band design, and the Fermi intensity level are dissimilar in the organic scenario. In addition to solitons, polarons, photoexcited, and non-linear optical absorption, other concepts such as solitons, polarons, and excitons, to name a few, must be properly considered to have a complete understanding of this newish classification of semiconductors (Miller and Abrahams, 1960; Gill, 1972). The above chapter concludes with an explanation of the components investigated in this study, including their chemical composition as well as optical properties.

2.2. MATERIALS AND THEIR CHEMICAL PROPERTIES

Large, heavy molecules result from the recurring of a fundamental unit known as a monomer to form polymer composites. OSC, as with all organic materials, are composed of carbon atoms, that also, in the scenario of polymers, constitute the basic chain. Attaching extra components or chemical bonding to the spine can alter its chemical components. Every three other carbon atoms to other carbon molecules and interacts with metal atoms through partial ionic interactions (Bässler, 1981; Gartstein and Conwell, 1995).

C's valence electron is He[2s22p2], that also indicates that two of the element's four electron pairs, the *p* electrons, are obtainable to shape a covalent bond. A really structure would result in the formation of two analogous bonds, even though carbons are capable of forming four. To clarify of that kind behavior, the concept of valence hybridization is applied, which postulates that one's particle is supported to the final unoccupied *p* orbital, resulting in four individually occupancy nucleon orbitals; then the remaining *s* as well as approximately or even all of the atomic orbitals (AOs) are combined to produce novel similar orbitals. The hybridized orbitals are thus a linear mixture of the original ones. The amount of *p* orbitals used during the hybridization procedure determines the kind of polymer as well as a number of its structural and chemical features (Novikov et al., 1998; Maennig et al., 2001). There are three potential hybridizations: sp^3, sp^2, and sp^1. The superscript displays the proportion of *p* orbitals participating in the hybrid. In Figure 2.1, a schematic demonstrates the hybridization idea.

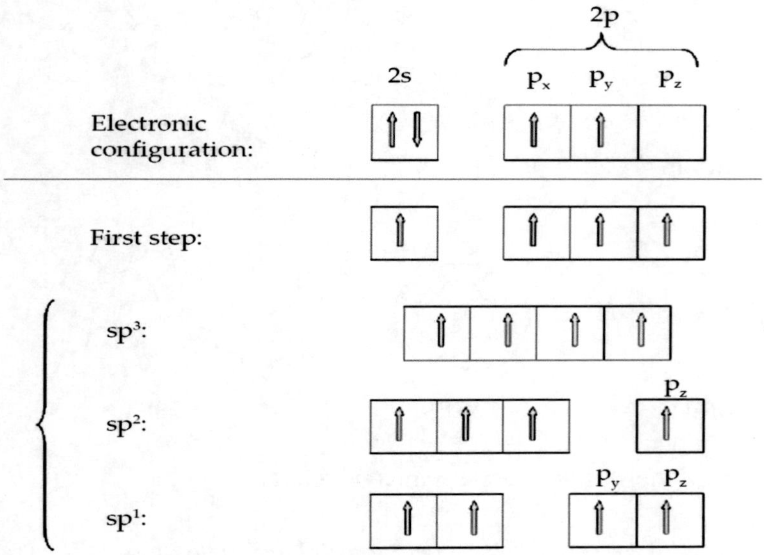

Figure 2.1. Valence orbital hybridization forms for the carbon atom.

Source: https://chem.libretexts.org/Ancillary_Materials/Reference/Organic_Chemistry_Glossary/Hybridization.

Between such two atoms, across each molecule's axis, the ⁻electron is dispersed. Significantly smaller than a covalent bond, the double molecule measures 1.34 compared to 1.54 for the single bond (Tanase et al., 2003).

When *sp*[1] hybridization is used, triple connections with a duration of 1.21 are formed, as seen in Figure 2.2 for the acetylene molecule.

Figure 2.2. Acetylene molecule.

Source: https://byjus.com/chemistry/acetylene-formula/.

Due to the alternating of single as well as double bonds throughout their spine, polymeric semiconductors with *sp*[2] hybridized carbon atoms are also known as conductive polymers. Likewise, the whole molecule of colorants is composed of cyclic components arranged in a network. In Figures 2.3 and 2.4, the structures of the following polymers and molecular substances are depicted.

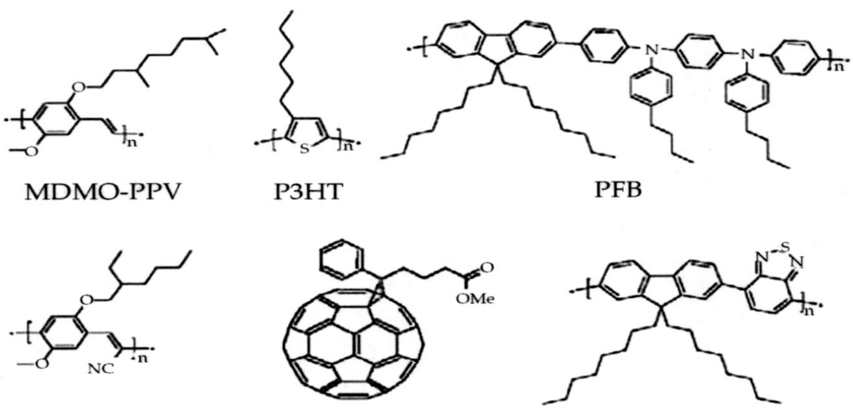

Figure 2.3. Models of polymeric semiconductors.

Source: https://www.researchgate.net/figure/Model-structures-1-7-corresponding-to-the-following-fragments-of-polymer-chains-initial_fig1_315660521.

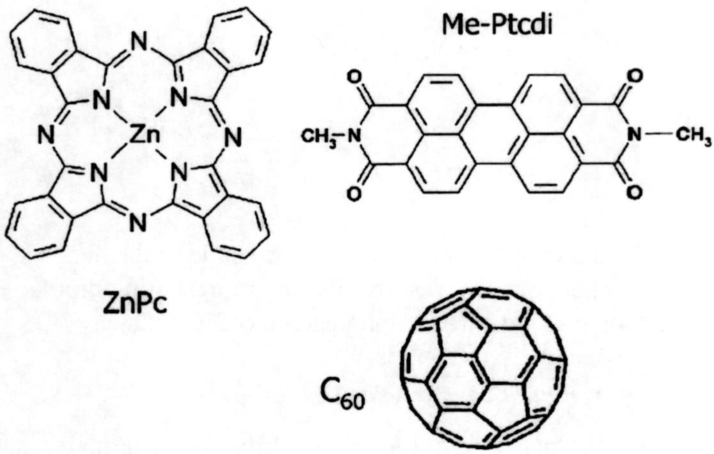

Figure 2.4. Instances of tiny molecule compounds.

Source: https://www.researchgate.net/figure/Chemical-structures-of-some-small-molecule-compounds-under-clinical-or-preclinical_fig2_344969526.

The crossover of every carbon atoms Pz wavefunction to that of its closest neighbor is what enables electron dissociation somewhere along the molecule's spine. PCBM ([6,6]-phenyl C61-butyric acid methyl ester) is an outlier amongst polymers, as shown in Figure 2.3 since it is composed of a C_{60} molecule (fullerene, in Figure 2.4) with an active site connected to the spheroid. The horizontal chain makes fullerene permeable in polar solvents, and in several circumstances, this substance is not appropriately classified as a polymer (Monroe, 1985; Juška et al., 2000).

Polymers and molecular substances share a delocalized extended-electron framework, which enables the transfer of photogenerated carriers (Takagi and Naito, 2021). The compound benzene, shown in Figure 2.5 in many ways similar, is a straightforward case of an electronic configuration.

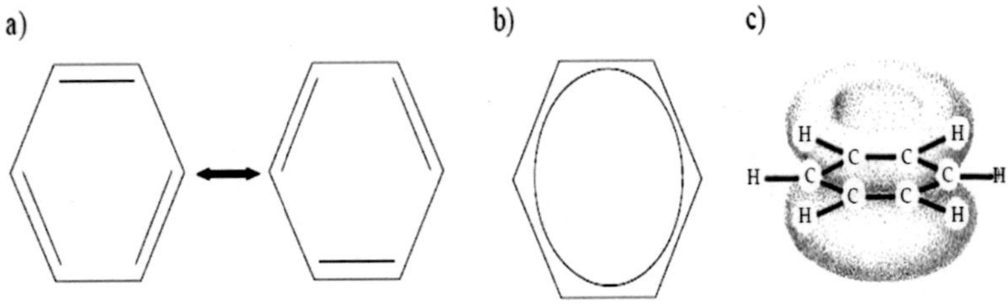

Figure 2.5. Various methodologies to depict the shape of the benzene molecule are shown: (a) resonance structures; (b) the ring representing dipolar electrons; and (c) the electronic cloud spread along each molecule plane.

Source: https://www.nature.com/articles/s42004-018–0057-4.

Because of the benzene molecule's ability to fluctuate across two similarly stable energy states, interchangeability between single as well as double links. Figure 2.7(b)'s structure depicting a circle continues graphically the dissociation of the six electrons that create an electronic cloud scattered along each molecular surface (Figure 2.7(c)) (Kasap et al., 1990; Tsung and So, 2008).

Similar to semiconducting materials, OSC are categorized into n-type and p-type groups. Nonetheless, as would be demonstrated in the following section, the doping methods utilized in organic compounds, such as silicon, are notably distinct from those used in inorganic molecules, such as silicon, so a real correlation with them at that stage might be incorrect. MDMO-PPV (poly[2-methoxy-5-(3,'7'-dimethyl-octyloxy)-p-phenylene vinylene]), P3HT (poly-3-hexylthiophene), and PFB (poly(9,90-dioctylfluorene-co-bis-N,NO-(4-butyl phenyl)-bis-N,NOphenyl-1, ZnPc (zinc phthalocyanine) is a p-type semiconductor, while C_{60} (fullerene) and Me-Ptcdi (N,N′-dimethylperylene-3,4,9,10-dicarboximide) are n-type semiconductors (Tiedje and Rose, 1981; Heun and Borsenberger, 1995).

Given the internal structures of such substances, it is feasible to determine the parameters required to produce an effective organic device:

- Conjugated network existence;
- Negative electron clouds in the material must be as overlapping as feasible;
- Excellent structural characteristics of thin films;
- Simple chemical pureness;
- Material integrity.

The first prerequisite has already been mentioned; the 2nd and 3rd conditions pertain to thin film qualities, the coating process, as well as its variables: the denser the structure, the more particle clouds intersect, and the more efficient the charge transfer. Likewise, the more organized a microstructure is the greater the thin film's electrical characteristics. In several instances, more material cleaning is required to get quality devices. In photovoltaic research and technology, material durability over exposure to environment and radiation is also a significant factor. Specifically, in the context of n-type transistors, the sensitivity of the exciting molecules produced in semiconductor materials might provide an issue (Tiedje, 1984; Naito and Kanemitsu, 1994).

2.3. BASIC WORKING PRINCIPLES

A few conceptual frameworks about semiconductor materials would be presented. The goal of this study is not to provide a possibly deep conversation on such a topic, but rather to provide an intuitive elaboration of the bandgap and main resonances in OSC. And although only dyes are being used in this task, polymers, as well as molecular components, will be shortly explained as a general overview. It is worth noting that several pieces of info are being debated, nonetheless, a certain text could be discovered that provides an emergence to such essential topics (Slowik and Chen, 1983; Ogawa and Naito, 2002).

On the one hand, (Ogawa, 2002) suggest categorizing OSC into three groups: polymers with the depraved lowest level, polymers with non-degenerate energy state, and specific molecular materials. The initial stable excitations (solitons, polarons, bipolarons, and excitons) change depending on the material (Borsenberger et al., 1996).

Another point of position on the matter contends that in certain kinds of natural semiconductors revealed to lighting, excitons are formed first and then disentangled into different electric charges (for instance polarons). The problem in analyzing a few experimental data has added to the confusion regarding what a critical difference is. Nonetheless, the charge generation mechanism via excitons formation is the most frequently cited in the literature devoted to the study reason on organic photovoltaics (OPVs) (Shen et al., 2003; Schein and Tyutnev, 2008).

2.3.1. Molecular Orbital Theory

In the existence of molecular systems, the molecular density of states model is used. While atoms or ions with the same energy communicate, their energies are divided, resulting in two various molecular energy states: one with an energy level than just the originals and the other with greater intensity. Since these small molecule vibrational modes are impossible to quantify precisely for most molecules, an estimation known as linear combination of atomic orbitals (LCAO) is utilized. As per LCAO, the molecular orbital closest to an atom could be regarded the isolated atoms (Bässler et al., 1994; Kobayashi et al., 2011). For two hydrogen atoms, the wave equation is as follows:

$$\Psi_{\pm} = \Psi_{1s}(A) \pm \Psi_{1s}(B)$$

$$with\,\Psi_{1s}(A) = \sqrt{\frac{1}{\pi a_0^3} \cdot e^{-r_A/a_0}}$$

wherein; A and B are atoms; and r_A is the range among electrons as well as A. The wavefunction for atom B is (2.2), but with RB representing the range between particle and molecule. Two molecular structures have been acquired by linearly incorporating the atomic wavefunctions in (2.1): the bonding orbital (+) and the anti-bonding orbital (–) (Bässler et al., 1994; Fishchuk et al., 2003).

The amplitudes of atomic and molecular orbitals are represented in Figure 2.6. The linking atomic orbital with an electronic structure greater than zero there between the two nuclei is obtained by adding the molecular nuclei. This arrangement gets the atoms together and is theoretically advantageous since its energy (E^+) is less than that of the solitary AOs. A gap between the orbitals, on the other hand, yields the anti-bonding electronic structure, with zero electronic structure in the center in between nuclei. Because its energy (E^-) is greater than that of the isolated atomic, this arrangement is unfavorable for molecule stabilization (Baranovskii et al., 1995; Oelerich et al., 2010).

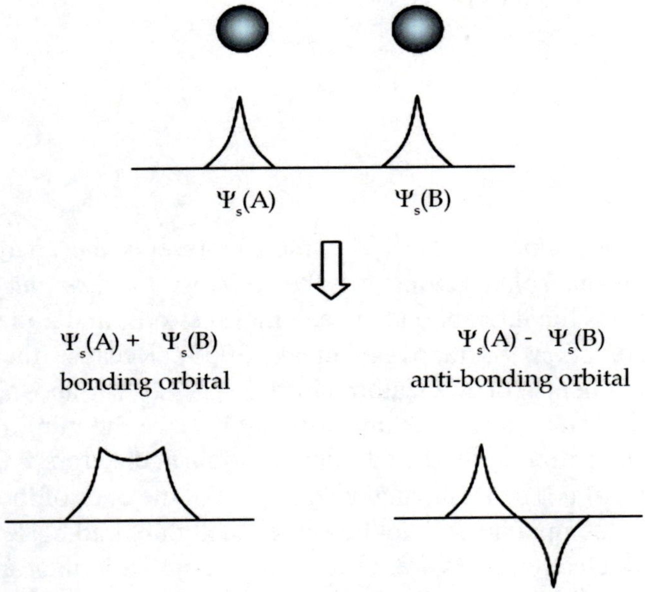

Figure 2.6. 3D depiction of atomic and wavefunction magnitude (top section in the diagram) and chemical subshells (lower part).

Source: https://chem.libretexts.org/Bookshelves/General_Chemistry/Map%3A_Chemistry_-_The_Central_Science_%28Brown_et_al.%29/06%3A_Electronic_Structure_of_Atoms/6.06%3A_3D_Representation_of_Orbitals.

The creation of molecular orbitals energy levels is seen in Figure 2.7.

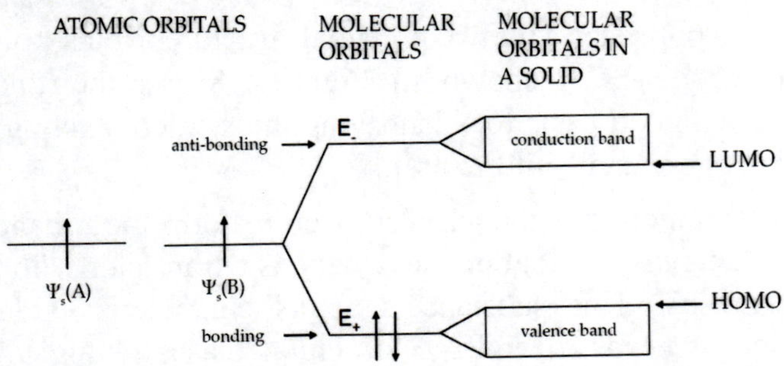

Figure 2.7. Levels of energy of two separated atoms, a diatomic molecule, and a solid.

Source: https://www.researchgate.net/figure/Representative-vibrational-energy-levels-and-rotation-of-a-diatomic-molecule-n-is-the_fig1_225490382.

The below are values of the two AOs:

$$E_+ = \frac{\beta - \gamma}{1 - \alpha}$$

$$E_- = \frac{\beta + \gamma}{1 + \alpha}$$

where β is the *coulombic integral*, which represents the energy that the electron has on the isolated atoms, γ is the *exchange (or resonance) integral*, which represents the interaction between the electrons, and α the amplitude of the wavefunction. (Baranovskii et al., 2000). Because there are two primary components (one per atom) in the case shown above, they both fill the bonding electronic structure, which will have the minimum energy, given opposing spinning. In the addition of a fourth electron, the molecular mass energy will not be less than the aggregate of the ones of the individual atoms, and the compound will not be energetically favorable. Helium atoms (He) with two electron pairs are, in fact, more durable in their atom. In the instance of a solid made of numerous molecules, the combination of all of the associating orbitals leads to a further dividing (see groups in Figure 2.7) along the emergence of bands. HOMO is the engaged molecular orbital (homo orbital in the conductive band, while LUMO is the lower energy level orbital in the bandgap. The energy gap is represented by the differential among HOMO and LUMO (Eg) (Lanyon, 1963; Hartenstein et al., 1996).

2.3.2. Polyacetylene: Band Structure and Primary Excitations

Polyacetylene is polyester with two potential structures, trans-polyacetylene and cis-polyacetylene, as shown in Figure 2.8. As the first polymer researched, it is a good basis for identifying the semiconducting properties of materials (Nagase and Naito, 2000).

Every carbon atom gives one *pz* electron to form the π-band. Though, experimental facts display that polyacetylene is a transistor with a disparity larger than 1.5 eV. This substance's actual characteristics lead to the conclusion that it needs to undergo structural biases owing to a process known as Peierls instability, which also involves transactivation of the molecule. The recurrence of the Peierls instability is $\Lambda = \pi/kF$ where kF is the Fermi wave vector. With a half-filled-band, $kF = \pi/2a$ as well as, due to the Peierls deformations, the basic unit is indeed not $(CH)_x$ but $(\text{-HC=CH-})_x$, indicating that the element is increased: dimerization. This dimerization is depicted diagrammatically as a variation of single as well as double bonds in Figure 2.7. The half-filled group is subdivided into a full-band, analogous to the conductive band in semiconductor materials, as well as a vacant-band, analogous to the bandgap (Suzuki et al., 2015; Uratani et al., 2016).

Overall, there are far more than two-bands, the amount of which depends on the number of carbon atoms in the repetition unit (Suzuki et al., 2015). In the case of polyacetylene, as a result of the dimerization we have two carbon atoms and two π-bands indicated as π and π*. To describe polyacetylene the SSH Hamiltonian, from Su, Schrieffer and Heeger theory, has been proposed starting from two assumptions:

- The electronic structure can be approximated using the tight-binding estimation, which indicates that the size of the bands is W = 2zt, where z is the carbon content that is the closest neighbor and t is the transmission essential. In the particular instance of polyacetylene, z equals 2, whereas t is assigned the value 2.5 eV. Consequently, and considering that there are two groups, the frequency range for both is W 5 eV (Levinson et al., 1982; Asada and Koseki, 2018).
- Thru the duration of the chemical bonded carbon atoms, the molecular sequence is çombined with the local electronic structure. Continuously switching between single as well as double bonds induces a structure-perturbing particle known as the soliton (Kamins and Marcoux, 1980).

The focus of this study is a detailed outline of the theory so that only the findings would be introduced here. A. J. Heeger and coworkers present a more comprehensive thesis. Constantly oscillating between both the two potential setups as well as around an equilibrium side chain value. This component is tried to introduce as a bouncing integral in the Hamiltonian as tries to follow:

$$t_{n,n+1} = t_0 + \alpha(u_{n+1} - u_n)$$

where; u_n is the dislocation of the nth carbon chain from its equilibrium point. The above word conveys the interplay between electrons as well as phonon, coupling the electronic properties to the topography of the particle. Consequently, the ultimate phase of the SSH Hamiltonian is as tries to follow:

$$H_{SSH} = \sum_{n,\sigma}[-t_0 + \alpha(u_{n+1} - u_n)](c^+_{n+1,\sigma}c_{n,\sigma} + c^+_{n,\sigma}c_{n+1,\sigma}) + \sum_n \frac{p_n^2}{2m} + \frac{1}{2}K\sum_n (u_{n+1} - u_n)^2$$

where; p_n are the nuclear momenta, u_n is the movements from balance, m is the mass of carbon; *K* is the efficient spring constant; as well as + *n*, *c*, and *n*, *c* are the fermion formation and destruction technicians for site n and spin. For $|u_n| > 0$, the system changes a sudden break in similarity, known as the Peierls distortion, as well as the amount of energy, is reduced (Sera et al., 1989; Angelis et al., 1999). Thus, we can approximate the mean-field as regards:

$$u_n \to \langle u_n \rangle (-1)\ u$$

Utilizing the equation, it can ascertain the value u_0 that reduces energy within this estimation (2.6). Evaluating that the energy is lessened in both u_0 as well as $-u_0$ attributes, two potential setups are depending on the role of double links, and the density of states of polyacetylene is two-fold degenerate (Yang et al., 2005).

The system's energy is visualized in Figure 2.8. The two points can be viewed as two equal-energy phases, period A as well as period B. The polymer is compelled to transform from one state to another by its stimulation. The soliton is the main excitation of polyacetylene (Zhang et al., 2009; Matsubara et al., 2011).

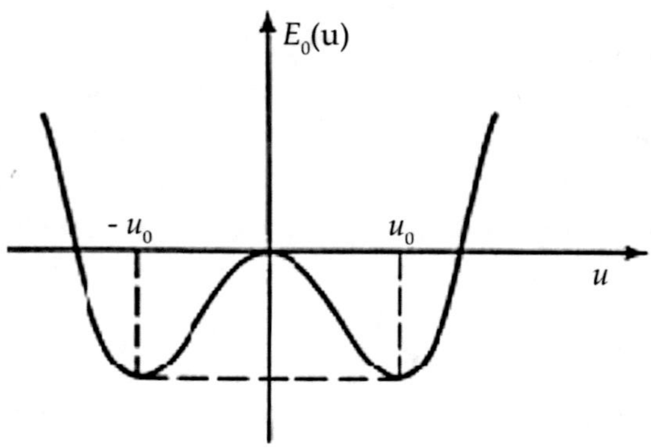

Figure 2.8. The system's potential concerning the dimerization position u.

Source: https://www.nature.com/articles/s41594-022-00746-2.

SSH theory forecasts that the structural relaxation among phases A and B ranges over a considerable distance.

As depicted above in this diagram, carbon atoms. However, for the sake of simplification, this component is often not considered when designing a soliton element system. With the creation of a soliton, a phase with energy would seem in the middle of the gap, that can be possessed by zero, one, or two electrons: with no electron, the soliton is positive charge and therefore has no twist; with one electron, the soliton has no charge and twist equitable to 12; and now with pair of electrons, the soliton is charged negatively and has no twist (Friedman, 1964; Kozlov et al., 2000).

Degradation in polyacetylene as a result of the linear similarity of the macromolecule. Essential to bear in mind is that the entrance of energy state into the space as a consequence of the creation of solitons is taken into consideration when determining the optical characteristics of the sample. After the production of solitons, additional transitions with sub-gap power are feasible: the near-infrared (NIR) region (Lu et al., 2020).

2.3.3. Polymers with Non-Degenerate Ground State

Polymers like poly(phenylene vinylene) (PPV) and poly para phenylene (PPP) had no degenerate equilibrium state. It is currently unknown whether the creatures produced by brightness are polarons as well as bipolarons straight or through the creation and detachment of consecutive excitons. Figure 2.8 depicts the structures of PPV and PPP; the degeneracy is divided because the two isoforms are not actively comparable (Root et al., 2017; Bagchi and Ediger, 2020).

As it has been explained earlier, the number of carbon atoms in the unit cell determines the number of bands for the π-electron system. The unit cell of PPV, for instance, contains 8 carbon atoms: 6 in the cyclic part of the structure and two in the other part (Figure 2.14-a). Such structure leads to have 8 π-sub-bands: 4 bonding and 4 anti-bonding ones. As every carbon atom contributes with 1 electron to the π-system, there are a total number of 8 electrons occupying the four bonding π-orbitals, while the other orbitals, characterized by a higher energy, are empty . Brazovskii et al. give a detailed description of the theory of electronic states in PPV (Brütting, 2005; Zhang et al., 2015).

A polaron is the bound state of a positively charged soliton and a negatively charged anti-soliton. Since there is only one stable phase (no degeneration), the soliton/anti-soliton (S-AS) couple does not result in a phase transition. Therefore, the soliton is counterbalanced by an anti-soliton, and the two energy levels originating from the two particles are separated into a bonding-less energetic level and an anti-bonding-more energetic level (Liu et al., 2013).

2.3.4. Small Molecules

Small molecules, an additional group of OSC, must be addressed in a separate area. Because their molecules are fairly tiny and geometric, colors and several dyes share certain characteristics. The Van der Waals force holds discrete molecules with each other to form molecular particulates. This weak force is the source of a few peculiar properties that differentiate molecular solids from others (metals, ionic solids, etc.). A general rule states that in the existence of a weak opponent, the stronger opponent will win despite the force holding molecules together, the characteristics of the individual particles are still preserved in the solid.

Excitons are the steady primary vibrational modes in molecular materials. Excitons are proven to occur in OSC and are consisting of a pair of electrons and holes that interact via the Coulomb force. As the total charge and spin of the particle are both zero, the exciton must be split to produce an electric current. There are three types of these particles: Frenkel, Wannier-Mott, and charge-transfer (CT) excitons. What distinguishes each form of the exciton is the separation among electron-hole pairs, i.e., the particle's radius: when it is on the sequence of the crystalline constant (aL), the exciton is of the Frenkel type; when the distance is greater, we have a CT exciton; for distances significantly greater than aL, we have Wannier-Mott excitons (Tokumoto et al., 1987; Henson et al., 2012).

The last form of the substance is the most prevalent in inorganic semiconductors, but it is only detectable at cold temperatures because, at ambient temperature, the thermal energy is greater than its binding energy and is sufficient to dissociate it. In polymers such as PDAs and organic molecular crystals (OMC), the steadiest excitons are Frenkel and CT excitons, and their bond length is approximately 0.5 eV. The stabilization of excitons in OSC with a low dielectric constant is dependent on the inability of the material to sufficiently separate two opposite charges. In a Frenkel exciton, the two carriers are located in the same molecule, whereas in a CT exciton, they are located in two separate molecules; however, they remain bound (Pecile et al., 1989; Bronstein et al., 2020).

As will be explained in the following chapter, the existence of excitons is a crucial consideration when constructing a solar cell from OSC. Similar to solitons, polarons, and bipolarons, energy levels exist in the gap when an OMC is illuminated and excitons are created. However, in this instance electron and hole are still coupled and wander arbitrarily along the molecule and from one molecule to the next: the exciton is unaffected by the electric field since it lacks charge. The presence of the exciton presupposes that the electron-hole connection is greater than the electron-phonon interaction that defines polarons, which is another intriguing element. Excitons are capable of motion, but charge carriers are not yet liberated (Jiang and Kloc, 2013; Martín et al., 2018).

To separate the exciton into a free electron and a free hole that may flow to their respective electrodes, a force is necessary. Explanation of the methods by which excitons are dissociated in a solar cell. In nature, the mechanism governing the activity of OSC is seen in the photosynthesis process, where photonic energy is extracted from incoming light and turned into chemical energy through a similar process (Zhu et al., 2008).

2.3.5. Experimental Proofs and an Alternative Model

The presence of the energized organisms discussed in the preceding paragraphs is supported by scientific proof. Thus, every atom is distinguished by getting a charge and/or a turn value and thus can be detected using various methods. Neutral resonances have no charge but a twist of 1/2. Electron spin resonance (ESR) and electron-nuclear double resonance (ENDOR) could even confirm the existence of spin in a substance. Sub-gap absorption spectra, as well as photoconductivity dimensions, as well as photoconductivity dimensions, can be used to investigate charged solitons with no spin. Solubility characteristics that can be seen in the NIR area of the spectrum are caused by the looks of energizing stages in the discrepancy. Photo-absorption results in the instant splitting of electron/hole into oppositely-charged particles, which also give rise to electric charge (Schwarze et al., 2016).

Charge storage could be investigated using a mixture of electrical, magnetic, and optical measurement techniques if the main pulses are polarons as well as bipolarons. Going to remember that polarons have a charge (|e|) as well as spin (1/2), whereas bipolarons have a charge (2|e|) but no spin, ESR, as well as optically detected magnetic resonance (ODMR), is used to distinguish between the various spin principles; optical measurements could be used to identify the directly applying to such particles in their spectrum. Detection of polarons and bipolarons has also been accomplished using optical readings (Brédas et al., 2002).

Eventually, in the existence of excitons, optical as well as photocurrent measured values can aid in differentiating of that kind electron-hole pairs from other atoms: the absorption spectrum onset is greater while tested in a photocurrent way than when evaluated purely optically. This is because of the neutral existence of excitons, which need further energy to detach and produce an electric current. The absorption initiation of PDAs (polyacetylenes), for instance, happens at 1.8–1.9 eV, an amount at that no photoconductivity is noticed. The photocatalytic onset occurs at 2.3–2.5 eV, which is high sufficient to tear electron-hole pairs that have been discovered throughout this way. Electroabsorption (EA) test results can be used to determine the excitons' activation affinity. The study of transient photocatalytic activity in the pico-, as well as nanosecond zones, is also a helpful approach (Tao et al., 2013).

Due to the large amount of experimental effort done to discover which processes govern the form of the energized pair in OSC, numerous questions remain. For instance, the true reasons for the onset mismatch between photocatalytic activity as well as optical measurements are currently unclear. Per a more simplistic perspective of the issue, action potentials are created under the light in even polymers or molecular substances, and they must then be dissociated into two independent electric charges in terms of achieving the flow of current (Newman et al., 2004).

Finally, in Table 2.1, a contrast of physicochemical nomenclature for exciting species formed in semiconductor materials is shown.

Table 2.1. Compares Chemistry and Physics Concepts: What a Physicist Calls an Atom is Referred to as an Exciting Reactive Species by a Pharmacist

Physical term	Chemical term
Neutral soliton	Neutral radical
Positive soliton	Spinless cation
Negative soliton	Spinless anion
Positive polaron (hole with associated lattice distortion)	Radical cation
Negative polaron (electron with associated lattice distortion)	Radical anion
Positive bipolaron	Spinless dication
Negative bipolaron	Spinless dianion

2.4. OPTICAL PROPERTIES

In photovoltaic (PV) technology, the optical qualities are of great relevance. The potential photophysical procedures in a conjugated particle or polymer. Based on the weak forces between molecules, of that kind procedures characterize the molecular scheme and are also available in a solid. All photophysical procedures that could happen in a conjugated particle or polymer while energized by light are depicted in the image in question. S_0, S_1, and S_n are the singlet conditions, whereas T_1 and T_n are the triplet states; nr k and' nr k is the long-term average for non-radiative procedures, r k and k are the constants for thermal procedures, and ISC k is the constant for the intersystem crossing (ISC) procedure. Each of these mechanisms competes and is distinguished by a decay time (the half-life, of the exciting organisms).

Traditional absorption and transmittance system perform the determination of the light intensity soaked up (or transmitted) in regards to optical gap transformation of the evaluated material. Absorption is only possible for photons with an energy amount greater than the energy gap (Eg). In the contaminates, new energy stages are introduced into the forbidden gap, necessitating more sensitive detection methods, such as photothermal deflection spectroscopy (PDS) (Zhang et al., 2017).

After the substance has consumed the light, it comes back to its initial state through a variety of methods. Photoluminescence (PL) phenomena are caused by the radiative decay products from singlet (S_1) and triplet (T_1) conditions to the initial state. The first process (S_1S_0) is recognized as fluorescence and has a decomposition time on the sequence of ns (10^{-9} s); the second method (T_1S_0) is recognized as phosphorescence and has a decomposition time on the sequence of s-ms (10^{-6}–10^{-3} s). This procedure necessitates a spin change, as the system must transition from the triplet state T_1 to the singlet state S_0. Transitions among states and regions with a variety of spins are significantly less likely than transitions among conditions with similar spins. PL procedures can be identified spectroscopically as time-dependent, to study the progression of the deterioration, or in stable mode, i.e., by calculating the total strength radiated over a specified time interval. Transfers between vibrational modes (S_nS_0 and T_nT_1) may be identified by Photoinduced Absorption Spectroscopy, for instance (PIA).

As will be discussed, organic solar cells need a heterojunction among a giver semiconductor and a receiver semiconductor. Figure 2.8 depicts the more intricate structure of probable photophysical activities in this instance.

Donor, as well as acceptor activity states, are shown alongside the energy state of charge-separated states (CSS). Transfers from one substance to the other are also feasible in a donor/acceptor system, resulting in a higher variety of photophysical phenomena. CSS are phases in which the electron, as well as hole, are split, resulting in two chargeable polarons that are free to travel to their electrode surface.

As previously mentioned, OSC are distinguished by nonlinear optical responses. As a result of the creation of excited atoms, the presence of an energy state in the space produces optical changes that were not there before absorbance. In less than a picosecond (10^{-12}), whenever photons strike, the midgap energy state related to the new electrons emerges. In Figure 2.8, a schematic depicts the aforementioned processes. Upon absorbing light, a band-to-band transition takes place, however, after 10^{-13} s, the system softens with the formation of a polaron, which is shown here as an illustration. New optical changes in the NIR red region range originate from the new energetic levels in the space, which are related to the polaron (for energy values inferior to the optical gap of the material). It is thus required to apply sub-picosecond time fixed characterization methods to comprehend which processes are operating shortly after absorption. This rapid relaxation of photoexcited carriers is crucial to the optical characteristics of semiconductor materials.

In OMCs, intragap phases are associated with the production of excitons. Excited singlet (S_1) and triplet (T_1) levels correlate to excitonic entities, the form among which (Frenkel or charge transport kind) should be studied in each instance (Newman et al., 2004). Owing to exciton production and band-to-band changes, the higher penetration component of the absorptivity of these materials consists of many characteristics.

2.5. TECHNOLOGICAL ASPECTS

2.5.1. Doping Mechanisms

As with doped silicon, OSC could be categorized as n-type and p-type materials. However, compared to the doping method typically used in the situation of silicon, this one is distinguishable. To create n-type and p-type semiconductors out of silicon, interstitial contaminants such as boron (B) or phosphorus (P) are introduced. A small number of silicon atoms are replaced with atoms from the other components to fulfill the requirements.

Given that a silicon atom has 4 outer electrons, *P* is an acceptor precipitate because it has five, whereas B is a donor precipitate. After all, it has three. Therefore, when silicon is doped with phosphorus, it transforms into an n-type semiconductor as well as new powerful stages linked to the pollutant seem so in the difference close to the bandgap; while silicon is doped with boron, it transforms into a p-type semiconductor and new energetic levels are introduced in the gap close to the valence band.

It is important to note that whenever discussing OSC, nomenclature borrowed from inorganic science is frequently misrepresentative and also that the processes by which they can be "doped" are distinct. Alternatively, acceptor (A) for the n-type and donor (D) for the p-type is used to define organic semiconductor materials. The terms "n-type" and "p-type" in the context of organic materials make the point that n-type semiconductor materials are highly conductive of electrons and "p-type" semiconductors are highly conductive of gaps. The recipient receives electrons from the donor in a solar cell (see paragraphs below).

In time to prevent terminology confusion, it is important to remember that while n-type silicon is produced by adding donor particulates, an organic n-type semiconductor is referred to as an electron carrier. Additionally, the act of "doping" an organic semiconductor, or adding electric charge to it, has the dual meaning of adding dipoles from a logical standpoint and generating chemically reactive organisms in the form of radical ions from a chemical perspective.

There are four ways to dope organic semiconductor materials:

- Chemical;
- Electrochemical;
- Photo-doping; and
- Interfacial.

The kind of molecules that are produced in a loaded organic semiconductor illustrates how a wide range of doping methods may exist. Even though charged particles could be established and described in several ways, this wide range is helpful when categorizing new technologies.

- **Chemical Doping:** Both oxidation (p-type doping) and reduction (n-type doping) responses are involved in that form of doping.

For polymers, the section demonstrates the processes:

$$p-type:(\pi-polymer)_n+\frac{3}{2}ny(I_2)\rightarrow[(\pi-polymer)^{+y}(I_3^-)_y]_n$$

$$n-type:(\pi-polymer)_n+\left\lfloor Na^+(Naphthalide)^{\cdot}\right\rfloor_y\rightarrow\left\lfloor(Na^+)_y(\pi-polymer)^{-y}\right\rfloor_n+(Naphth)^0$$

After which, it is necessary to form an electric charge in the particles chemically. As was previously demonstrated, chemists generate charged as well as reactive dissolved substances (radical ions), while physicists generate electric charge (polarons) helpful for electrical properties in application areas. Whenever the doping intensity is increased sufficiently, a material can change from conducting polymer to steel using this technique. High doping values are significantly simple to obtain. This method's weakness is that it is challenging to regulate, trying to make it frequently hard to attain uniformly and transitional stages of doping.

- **Electrochemical Doping:** The electrochemical method is frequently more practical given the difficulty in attempting to control the electron density that influences the chemical means. The semiconductor receives a redox charge from a diode, and the automated charge is balanced by ions diffusing into or out of the microelectronics from the neighboring analyte. Controlling the doping level is possible thanks to the voltage among the polymer and the electrode. The example below shows how the process works:

$$p-type:(\pi-polymer)_n+[Li^+(BF_4^-)]_{soln}\rightarrow[(\pi-polymer)^{+y}(BF_4^-)_y]_n+Li(electrode)$$

$$n-type:(\pi-polymer)_n+Li(electrode)\rightarrow\left\lfloor(Li^+)_y(\pi-polymer)^{-y}\right\rfloor_n+[Li^+(BF_4^-)]_{soln}$$

- **Photo-Doping:** The substance is locally oxidized and nearby reduced as a consequence of the synthesis of an electron-hole pair by photoabsorption and the following interfacial charge that allows for the development of divided electric charge. The following expression may be used to restart the method:

$$(\pi-polymer)_n+hv\rightarrow\left\lfloor(\pi-polymer)^{+y}+(\pi-polymer)^{-y}\right\rfloor_n$$

where; *n* represents how many electron-hole couples there are. The question of whether photo-absorption generates free agents or constrained excitons is still up for debate (Zhang et al., 2017) and appears to be dependent on the substance. In summary, it has to be tested out individually.

Numerous photo-physical mechanisms fight for control of the material's reaction after absorbing light. Coupling a donor as well as an acceptor may improve electron transport as well as the resultant creation of a charge carrier (see paragraphs below). The picture leads to the formation of both electric charges as well as necessitates a device to segregate them, whereas chemical processes enable charge transport creation of the desired signal.

- **Interfacial Carriers Injection:** At the point where a metal connection meets a semiconductor, charged particles may be introduced.

The given equations show that charges are immediately injected into π and π* bands:

$$p-type: \qquad (\pi - polymer)_n - y(e^-) \rightarrow \left\lfloor (\pi - polymer)^{+y} \right\rfloor_n$$

$$n-type: \qquad (\pi - polymer)_n + y(e^-) \rightarrow \left\lfloor (\pi - polymer)^{-y} \right\rfloor_n$$

where *y* is the number of injected particles. Equation refers to the injection of a hole into an otherwise filled π-band, or the extraction of an electron, which corresponds to an oxidation reaction. Equation refers to the injection of an electron into an empty π*-band, which consists in a reduction of the material.

The many processes mentioned above cause the substances to behave in various ways. Photo-current doping is constrained by decay and mixing processes, but electrochemical reactions doping's team is led is constant till the ions are balanced. Last but not least, in the instance of interfacial charge injection, the existence of a bias provided to the contacts limits the longevity of the charged electrons.

REFERENCES

1. Angelis, C. T., Dimitriadis, C. A., Miyasaka, M., Farmakis, F. V., Kamarinos, G., Brini, J., & Stoemenos, J., (1999). Effect of excimer laser annealing on the structural and electrical properties of polycrystalline silicon thin-film transistors. *Journal of Applied Physics, 86*(8), 4600–4606.
2. Asada, T., & Koseki, S., (2018). Simulation study of hole mobility in the amorphous phase of organic molecules. *Organic Electronics, 53*, 141–150.
3. Bagchi, K., & Ediger, M. D., (2020). Controlling structure and properties of vapor-deposited glasses of organic semiconductors: Recent advances and challenges. *The Journal of Physical Chemistry Letters, 11*(17), 6935–6945.
4. Baranovskii, S. D., Cordes, H., Hensel, F., & Leising, G., (2000). Charge-carrier transport in disordered organic solids. *Physical Review B, 62*(12), 7934.
5. Baranovskii, S. D., Faber, T., Hensel, F., & Thomas, P., (1997). The applicability of the transport-energy concept to various disordered materials. *Journal of Physics: Condensed Matter, 9*(13), 2699.
6. Baranovskii, S. D., Thomas, P., & Adriaenssens, G. J., (1995). The concept of transport energy and its application to steady-state photoconductivity in amorphous silicon. *Journal of Non-Crystalline Solids, 190*(3), 283–287.
7. Baranovskii, S. D., Zhu, M., Faber, T., Hensel, F., Thomas, P., Von, D. L. M. B., & Van, D. W. W. F., (1997). Thermally stimulated conductivity in disordered semiconductors at low temperatures. *Physical Review B, 55*(24), 16226.
8. Bässler, H., (1981). Localized states and electronic transport in single component organic solids with diagonal disorder. *Physica Status Solidi (b), 107*(1), 9–54.
9. Bässler, H., (1993). Charge transport in disordered organic photoconductors. A Monte Carlo simulation study. *Physica Status Solidi B (Basic Research); (Germany), 175*(1).
10. Bässler, H., Borsenberger, P. M., & Perry, R. J., (1994). Charge transport in poly (methylphenylsilane): The case for superimposed disorder and polaron effects. *Journal of Polymer Science Part B: Polymer Physics, 32*(9), 1677–1685.

11. Blom, P. W., (2020). Polymer electronics: To be or not to be?. *Advanced Materials Technologies, 5*(6), 2000144.
12. Borsenberger, P. M., & Bässler, H., (1991). Concerning the role of dipolar disorder on charge transport in molecularly doped polymers. *The Journal of Chemical Physics, 95*(7), 5327–5331.
13. Borsenberger, P. M., & Bässler, H., (1992). The role of polar additives on charge transport in molecularly doped polymers. *Physica Status Solidi (b), 170*(1), 291–302.
14. Borsenberger, P. M., & O'Regan, M. B., (1995). The role of dipole moments on hole transport in triphenylamine doped poly (styrene). *Chemical Physics, 200*(1, 2), 257–263.
15. Borsenberger, P. M., Gruenbaum, W. T., & Magin, E. H., (1996). Hole transport in vapor deposited bis (ditolylaminostyryl) benzene. *Physica B: Condensed Matter, 228*(3, 4), 226–232.
16. Brédas, J. L., Calbert, J. P., Da Silva, F. D. A., & Cornil, J., (2002). Organic semiconductors: A theoretical characterization of the basic parameters governing charge transport. *Proceedings of the National Academy of Sciences, 99*(9), 5804–5809.
17. Bronstein, H., Nielsen, C. B., Schroeder, B. C., & McCulloch, I., (2020). The role of chemical design in the performance of organic semiconductors. *Nature Reviews Chemistry, 4*(2), 66–77.
18. Brütting, W., (2005). Introduction to the physics of organic semiconductors. *Physics of Organic Semiconductors,* 1–14.
19. Fishchuk, I. I., Kadashchuk, A., Bässler, H., & Nešpůrek, S., (2003). Nondispersive polaron transport in disordered organic solids. *Physical Review B, 67*(22), 224303.
20. Friedman, L., (1964). Transport properties of organic semiconductors. *Physical Review, 133*(6A), A1668.
21. Gartstein, Y. N., & Conwell, E. M., (1995). High-field hopping mobility in molecular systems with spatially correlated energetic disorder. *Chemical Physics Letters, 245*(4, 5), 351–358.
22. Gill, W. D., (1972). Drift mobilities in amorphous charge-transfer complexes of trinitrofluorenone and poly-n-vinylcarbazole. *Journal of Applied Physics, 43*(12), 5033–5040.
23. Hartenstein, B., Bässler, H., Jakobs, A., & Kehr, K. W., (1996). Comparison between multiple trapping and multiple hopping transport in a random medium. *Physical Review B, 54*(12), 8574.

24. Henson, Z. B., Müllen, K., & Bazan, G. C., (2012). Design strategies for organic semiconductors beyond the molecular formula. *Nature Chemistry, 4*(9), 699–704.

25. Heun, S., & Borsenberger, P. M., (1995). A comparative study of hole transport in vapor-deposited molecular glasses of N, N′, N ″, N‴-tetrakis (4-methylphenyl)-(1, 1′-biphenyl)-4, 4′-diamine and N, N′-diphenyl-N, N′-bis (3-methylphenyl)-(1, 1′-biphenyl)-4, 4′-diamine. *Chemical Physics, 200*(1, 2), 245–255.

26. Hirao, A., & Nishizawa, H., (1996). Measurement of diffusion and drift of charge carriers from photocurrent transients. *Physical Review B, 54*(7), 4755.

27. Jiang, H., & Kloc, C., (2013). Single-crystal growth of organic semiconductors. *MRS Bulletin, 38*(1), 28–33.

28. Jiang, W., Li, Y., & Wang, Z., (2013). Heteroarenes as high performance organic semiconductors. *Chemical Society Reviews, 42*(14), 6113–6127.

29. Juška, G., Arlauskas, K., Viliūnas, M., Genevičius, K., Österbacka, R., & Stubb, H., (2000). Charge transport in π-conjugated polymers from extraction current transients. *Physical Review B, 62*(24), R16235.

30. Kamins, T. I., & Marcoux, P. J., (1980). Hydrogenation of transistors fabricated in polycrystalline-silicon films. *IEEE Electron Device Letters, 1*(8), 159–161.

31. Kasap, S. O., Polischuk, B., & Dodds, D., (1990). An interrupted field time-of-flight (IFTOF) technique in transient photoconductivity measurements. *Review of Scientific Instruments, 61*(8), 2080–2087.

32. Kobayashi, T., Kinoshita, K., Nagase, T., & Naito, H., (2011). Continuous-wave photoinduced absorption studies in polythiophene and fullerene blended thin films. *Physical Review B, 83*(3), 035305.

33. Kozlov, V. G., Parthasarathy, G., Burrows, P. E., Khalfin, V. B., Wang, J., Chou, S. Y., & Forrest, S. R., (2000). Structures for organic diode lasers and optical properties of organic semiconductors under intense optical and electrical excitations. *IEEE Journal of Quantum Electronics, 36*(1), 18–26.

34. Lanyon, H. P. D., (1963). Electrical and optical properties of vitreous selenium. *Physical Review, 130*(1), 134.

35. Levinson, J., Shepherd, F. R., Scanlon, P. J., Westwood, W. D., Este, G., & Rider, M., (1982). Conductivity behavior in polycrystalline

semiconductor thin film transistors. *Journal of Applied Physics, 53*(2), 1193–1202.

36. Li, R., Hu, W., Liu, Y., & Zhu, D., (2010). Micro-and nanocrystals of organic semiconductors. *Accounts of Chemical Research, 43*(4), 529–540.
37. Liu, X., Jeong, K. S., Williams, B. P., Vakhshouri, K., Guo, C., Han, K., & Asbury, J. B., (2013). Tuning the dielectric properties of organic semiconductors via salt doping. *The Journal of Physical Chemistry B, 117*(49), 15866–15874.
38. Lu, C., Liu, Q., Sun, Q., Hsieh, C. Y., Zhang, S., Shi, L., & Lee, C. K., (2020). Deep learning for optoelectronic properties of organic semiconductors. *The Journal of Physical Chemistry C, 124*(13), 7048–7060.
39. Maennig, B., Pfeiffer, M., Nollau, A., Zhou, X., Leo, K., & Simon, P., (2001). Controlled p-type doping of polycrystalline and amorphous organic layers: Self-consistent description of conductivity and field-effect mobility by a microscopic percolation model. *Physical Review B, 64*(19), 195208.
40. Martín, J., Dyson, M., Reid, O. G., Li, R., Nogales, A., Smilgies, D. M., & Stingelin, N., (2018). On the effect of confinement on the structure and properties of small-molecular organic semiconductors. *Advanced Electronic Materials, 4*(1), 1700308.
41. Matsubara, R., Sakai, M., Kudo, K., Yoshimoto, N., Hirosawa, I., & Nakamura, M., (2011). Crystal order in pentacene thin films grown on SiO2 and its influence on electronic band structure. *Organic Electronics, 12*(1), 195–201.
42. Miller, A., & Abrahams, E., (1960). Impurity conduction at low concentrations. *Physical Review, 120*(3), 745.
43. Monroe, D., (1985). Hopping in exponential band tails. *Physical Review Letters, 54*(2), 146.
44. Nagase, T., & Naito, H., (2000). Localized-state distributions in molecularly doped polymers determined from time-of-flight transient photocurrent. *Journal of Applied Physics, 88*(1), 252–259.
45. Nagase, T., Kishimoto, K. H., & Naito, H., (1999). High resolution measurement of localized-state distributions from transient photoconductivity in amorphous and polymeric semiconductors. *Journal of Applied Physics, 86*(9), 5026–5035.

46. Naito, H., & Kanemitsu, Y., (1994). Relations between transient charge transport and the glass-transition temperature in amorphous chalcogenides. *Physical Review B, 49*(15), 10131.
47. Newman, C. R., Frisbie, C. D., Da Silva, F. D. A., Brédas, J. L., Ewbank, P. C., & Mann, K. R., (2004). Introduction to organic thin film transistors and design of n-channel organic semiconductors. *Chemistry of Materials, 16*(23), 4436–4451.
48. Novikov, S. V., Dunlap, D. H., Kenkre, V. M., Parris, P. E., & Vannikov, A. V., (1998). Essential role of correlations in governing charge transport in disordered organic materials. *Physical Review Letters, 81*(20), 4472.
49. Oelerich, J. O., Huemmer, D., Weseloh, M., & Baranovskii, S. D., (2010). Concentration dependence of the transport energy level for charge carriers in organic semiconductors. *Applied Physics Letters, 97*(14), 214.
50. Ogawa, N., & Naito, H., (2002). Transient hopping transport in percolation clusters. *Electrical Engineering in Japan, 140*(2), 1–9.
51. Pecile, C., Palnelli, A., & Girlando, A., (1989). Studies of organic semiconductors for 40 years—V. *Molecular Crystals and Liquid Crystals, 171*(1), 69–87.
52. Root, S. E., Savagatrup, S., Pais, C. J., Arya, G., & Lipomi, D. J., (2016). Predicting the mechanical properties of organic semiconductors using coarse-grained molecular dynamics simulations. *Macromolecules, 49*(7), 2886–2894.
53. Root, S. E., Savagatrup, S., Printz, A. D., Rodriquez, D., & Lipomi, D. J., (2017). Mechanical properties of organic semiconductors for stretchable, highly flexible, and mechanically robust electronics. *Chemical Reviews, 117*(9), 6467–6499.
54. Schein, L. B., & Tyutnev, A., (2008). The contribution of energetic disorder to charge transport in molecularly doped polymers. *The Journal of Physical Chemistry C, 112*(19), 7295–7308.
55. Schwarze, M., Tress, W., Beyer, B., Gao, F., Scholz, R., Poelking, C., & Leo, K., (2016). Band structure engineering in organic semiconductors. *Science, 352*(6292), 1446–1449.
56. Sera, K., Okumura, F., Uchida, H., Itoh, S., Kaneko, S., & Hotta, K., (1989). High-performance TFTs fabricated by XeCl excimer laser annealing of hydrogenated amorphous-silicon film. *IEEE Transactions on Electron Devices, 36*(12), 2868–2872.

57. Shen, Y., Diest, K., Wong, M. H., Hsieh, B. R., Dunlap, D. H., & Malliaras, G. G., (2003). Charge transport in doped organic semiconductors. *Physical Review B, 68*(8), 081204.
58. Slowik, J. H., & Chen, I., (1983). Effect of molecular rotation upon charge transport between disordered carbazole units. *Journal of Applied Physics, 54*(8), 4467–4473.
59. Suzuki, F., Shizu, K., Kawaguchi, H., Furukawa, S., Sato, T., Tanaka, K., & Kaji, H., (2015). Multiscale simulation of charge transport in a host material, N, N′-dicarbazole-3, 5-benzene (mCP), for organic light-emitting diodes. *Journal of Materials Chemistry C, 3*(21), 5549–5555.
60. Takagi, K., & Naito, H., (2021). Characterization of transport properties of organic semiconductors using impedance spectroscopy. *Organic Semiconductors for Optoelectronics,* 137–159.
61. Takimiya, K., Shinamura, S., Osaka, I., & Miyazaki, E., (2011). Thienoacene-based organic semiconductors. *Advanced Materials, 23*(38), 4347–4370.
62. Tanase, C., Meijer, E. J., Blom, P. W. M., & De Leeuw, D. M., (2003). Unification of the hole transport in polymeric field-effect transistors and light-emitting diodes. *Physical Review Letters, 91*(21), 216601.
63. Tao, Y., Xiao, J., Zheng, C., Zhang, Z., Yan, M., Chen, R., & Huang, W., (2013). Dynamically adaptive characteristics of resonance variation for selectively enhancing electrical performance of organic semiconductors. *Angewandte Chemie, 125*(40), 10685–10689.
64. Tiedje, T., & Rose, A., (1981). A physical interpretation of dispersive transport in disordered semiconductors. *Solid State Communications, 37*(1), 49–52.
65. Tiedje, T., (1984). Time-resolved charge transport in hydrogenated amorphous silicon. *The Physics of Hydrogenated Amorphous Silicon II,* 261–300.
66. Tokumoto, M., Anzai, H., Ishiguro, T., Saito, G., Kobayashi, H., Kato, R., & Kobayashi, A., (1987). Electrical and magnetic properties of organic semiconductors, $(BEDT\text{-}TTF)_2X$ (X= IBr_2, IBrCl, and ICl_2). *Synthetic Metals, 19*(1–3), 215–220.
67. Tsung, K. K., & So, S. K., (2008). Carrier trapping and scattering in amorphous organic hole transporter. *Applied Physics Letters, 92*(10), 95.

68. Uratani, H., Kubo, S., Shizu, K., Suzuki, F., Fukushima, T., & Kaji, H., (2016). Detailed analysis of charge transport in amorphous organic thin layer by multiscale simulation without any adjustable parameters. *Scientific Reports, 6*(1), 1–7.

69. Yang, H., Shin, T. J., Ling, M. M., Cho, K., Ryu, C. Y., & Bao, Z., (2005). Conducting AFM and 2D GIXD studies on pentacene thin films. *Journal of the American Chemical Society, 127*(33), 11542–11543.

70. Zhang, F., Zang, Y., Huang, D., Di, C. A., Gao, X., Sirringhaus, H., & Zhu, D., (2015). Modulated thermoelectric properties of organic semiconductors using field-effect transistors. *Advanced Functional Materials, 25*(20), 3004–3012.

71. Zhang, J., Xu, W., Sheng, P., Zhao, G., & Zhu, D., (2017). Organic donor–acceptor complexes as novel organic semiconductors. *Accounts of Chemical Research, 50*(7), 1654–1662.

72. Zhang, X. H., Tiwari, S. P., Kim, S. J., & Kippelen, B., (2009). Low-voltage pentacene organic field-effect transistors with high-κ HfO_2 gate dielectrics and high stability under bias stress. *Applied Physics Letters, 95*(22), 312.

73. Zhu, Y., Kulkarni, A. P., Wu, P. T., & Jenekhe, S. A., (2008). New ambipolar organic semiconductors. 1. Synthesis, single-crystal structures, redox properties, and photophysics of phenoxazine-based donor−acceptor molecules. *Chemistry of Materials, 20*(13), 4200–4211.

CHAPTER 3

ORGANIC SEMICONDUCTORS FOR DEVICE APPLICATIONS

CONTENTS

3.1. INTRODUCTION

An extensive variety of organic semiconductors (OSC) have been presently conducting extensive research aimed at their possible uses through organic electronics (OEs). This is mainly owing to the increasing need to replace silicon (Si) with just some alternate solutions cost-effective components contribute small and simple as well as economically feasible techniques (Dodabalapur et al., 1995). At least in certain specialized areas, this trend is driving the investigation of a variety of OSC. The process that involves the publishing of semiconductor, dielectric material, and dielectric trends on a wide range of hard but also various substrates are developing rapidly as part of a financially successful software for the forthcoming OE devices. These advancements are the result of recent advancements in the field of operating systems (OSC) (Dodabalapur et al., 1995; Friend et al., 1999).

In way of comparison to OSC, which is primarily comprised of silicon, the extra potential of chemically changing organic compounds through the incorporation of a variety of functional organizations during synthesized results in several chemical features and functionality in addition to increased impact on the efficient charge properties (Sirringhaus et al., 1999). This is going to add even further impetus to start making that kind of search extra attractive and impactful. The cases that will be discussed subsequently provide even more light on just this idea. For example, it is now simple to construct molecules to get the characteristics that are wanted, such as their accessibility in a particular solvent, the hue of the emission of light, and a particular crystalline molecule stacking, to mention a few of these characteristics. Some of these capabilities, which were achieved by changing the molecular plans, are previously placed to be used in a variety of newer technologies, as will be shown in the following paragraphs (Kagan et al., 1999; Crone et al., 2000).

An excellent case in point would be that of nonvolatilizable reminiscence components (Coe et al., 2002), in which polyvinylidene fluoride copolymers and tri/tetra-fluoroethylene radical (Malliaras and Friend, 2005) were utilized in the process of manufacturing helpful devices that have been discovered to be particularly suitable for stretchy circuit boards. In a different scenario, the end units of an organically sensing device were changed in such a way that they not just to reacted to biological and chemical organisms but also detected pH levels (Popinciuc et al., 2006), the freshness of food, hazardous substances, tension, and stress in apparel and accessories (Naber et al., 2005; Guo et al., 2006) . The rapid expansion that is being seen in OE is providing numerous novel application areas that have superior performances, increased reliability and stability, extended lifetimes, excellent control, and repeatability following the requirements of the various industrial divisions. For example, an organic light-emitting device (OLED)-based displays, which are advertised by Philips as well as Carbon-based Electronic Elements and can be found in mobile devices such as cellular phones as well as digital phones, and car radios and image sensors, are among the application areas that are anticipated to increase in number in the coming years (Gundlach et al., 1999; Guo et al., 2006).

Within the recently period, the desire to explore the electrical applications for OSC already has developed a manifold, and it is being done in conjunction with an endeavor to eliminate the usage of inorganic compounds in several key fields. In addition, work is being done to investigate more recent ideas and theoretical frameworks, such as designs at the molecular scale, to manipulate the structural, physiological, as well as chemical possessions of natural molecules to more effectively meet the required functionality in the foreseeable future. Because previous attempts made at recognizing OSC devices were met with the unavoidable issue of substance instabilities during computation, in addition to the large concentration of structural deficiencies presented throughout physical growth, which prevented the broader misuse of both the intrinsic efficient care of physical possessions, because that kind of strategy is presently more relevant (Meijer et al., 2003).

OSC is presently growing close to being realistically ready for augmenting and/or replacing the traditional OSC, particularly in certain specialized sectors as a result of the specific benefits that are linked with this transition. Since the 1980s, OSC have gone through a series of phases of growth intended to improve material quality. These stages led to OSC reaching and smooth surpassing the achievement of amorphous silicon (a-Si) (Kline and McGehee, 2003) in aspects of the charge transport mobilities in organic thin-film transistors (OTFTs). Simpler result publishing of the natural product at low-temperature changes on even equitably large geographical substrata instead of using a vacuum environment system is very well worth consideration. Whereas processes which include screen, ink-jet, as well as microcontact productions have indeed been particularly developed for flexible device falsifications going to employ plastic polymers offering equally plausible assimilation of LEDs and organic photovoltaic (OPV) components, it is important to note that these (Zhang et al., 2003; Ahmad, 2014).

A solitary semiconductor is just a perfect platform for having studied and trying to exploit the innate efficient charge ionic conductivity, so comprehensive efforts have been put in now to develop crystalline OSC, and consecutive proper approval in content crystallization showed substantial improvement in the charge transfer. For instance, relatively low levels of the mobilities, observed in previous examples, within the range of 0.001 to 0.1 cm^2/Vs, were increased to as much as 20 cm^2/Vs in crystallized rubrene and commonly known samples at ambient temperature. At a temperature of 10 K, naphthalene isolates were found to have the mobility that was much greater than before, measuring 400 cm^2/Vs. Because of this sort of enhancement in electron mobility, there was a significant push toward employing them in device falsifications (De Boer et al., 2003; Lang et al., 2004).

In addition to being simple to manufacture, organic semiconductor-based strategies, such as organic light-emitting diodes (OLEDs), OPV solar cells, and organic field-effect transistors (OFETs), are rapidly being created. These devices make use of covalently organic compounds, which offer unique properties such as tunable energy band-gaps, redox possibilities, and charge transportation transit. The use of these techniques and materials will undoubtedly result in the production of lightweight, low-cost, thin-film, huge, and flexible and stretchable devices in the future (Sundar et al., 2004; Chua et al., 2005).

3.2. ORGANIC MOLECULES FOR DEVICE APPLICATIONS

When it comes to technological applications, it is useful to separate organic compounds into groups based on the types of charge carrier transport that they endorse, such as hole and electron-transporting (HT/ET) materials, in addition to their building, which includes small-molecule oligomers, macromolecular polymeric materials, and dendrimers (Gelinck et al., 2004).

The concept of p- or n-type materials that are often employed does not always correspond to the inherent capacity of such a carbon-based physical to carry hovels or electrons as described in scholarly studies. As opposed to that, it just describes how readily hovels and electrons are pumped from respective connection electrodes. This subtle departure from the traditional definition would be further developed in later studies, wherever it was demonstrated that while the inherent electron and hole vibrations could be comparable in very many organic compounds, their substantially decreased values, evaluated experiments, could be the result of multiple influences of hooks or instabilities presented by water, hydroxyl groups, or oxidation exposures. The SiO_2 gate dielectric's highly dense hydroxyl functional groups on the surface, that served as traps for electron transition into the channels, served as another example of how the environment may affect intrinsic charge transport mobilities. However, by coating the dielectric with BCB, substances including polythiophene, polyfluorene, and polyparaphenylene-vinylene, have efficient responsibility electron mobility in the range of 10^{-2} to 10^{-3} cm^2/Vs, were able to achieve excellent transfer of electrons (Rogers and Katz, 1999; Dimitrakopoulos and Malenfant, 2002).

Aromatic hydrocarbons, merged heterocyclic aromatics, oligothiophenes, oligoarylenes, macrocycles like phthalocyanines, fullerenes, and perylene pigments, as well as perylene and violanthrone as electron donors and formative research, are examples of biological molecules used during technological applications. The most effective compounds for OE to date are rubrene, tetracene, and crystalline pentacene. The attention to oligothiophenes has already been rapidly increasing since the invention of sexithienyl-based organic transistors. The fluorinated backbones of oligothiophenes and oligoacenes do exhibit n-type behavior even though they are mostly p-type materials (Horowitz, 1999). TPD from the triphenylamine (TPA) family that were vacuum-deposited was widely employed as a hole-transporting layer (HTLs) in OLEDs. When perylene tetracarboxylic dianhydride (PTCDA) or perylene tetracarboxylic diimide (PTCDI) moieties were attached, n-type materials were formed, and perylene displayed a herringbone-like molecular packing (Murphy and Frechet, 2007).

OFETs are routinely made using pentacene and a variety of many other aromatic hydrocarbons. Presently, raw along with alkyl substituted versions employing 16-, 20-, and 27-mer oligothiophenes are now being studied for optoelectronic applications, particularly in the form of molecular wires. Amorphous oligothiophenes also have been employed in numerous strategies in addition to crystallized oligothiophenes, according to published literature (Bao et al., 1997; Mas-Torrent and Rovira, 2008).

3.3. MATERIAL SELECTION CRITERIA FOR OSC DEVICES

Early advances of OFETs used the straightforward substitution of an organic channel for an artificial one based on the theory of a regular FET. To create a device, the designed source/drain connections on a severely doped Si substrate are covered with an organic phase. The Si gate then produces an electric charge in the initial section following the positive gate bias by altering the channel conductance (Chang et al., 1998).

Even though the OFET channel differs significantly from standard Si, the fundamental transistor design and operation are the same as their inorganic counterparts. In later advances, organic gate electrodes and insulating layers were being used to create all-organic systems, but these devices were shown to be noticeably less effective. A crucial FET characteristic is the charge carrier movement within the channel area because the greater the efficient charge movement, the quicker the device changes. As a result, achieving the highest mobility and on/off ratios is the main goal of device design and production. As a benchmark for contrasting the new OSC devices, a–Si FET's usual mobility and on/off conversion rate, for instance, are around 1 cm^2/Vs and 108, correspondingly (Hebner et al., 1998).

The charge transporters in such a carbon-based device "hop" through one–orbital to something on the backbone of the small molecule despite having the very same device architectural style and electrical operational processes as such an inorganic device. Numerous techniques are developed to increase portability and on/off ratio, as will be briefly discussed in following section. The alignment of the energy state of the contacting electrodes, as well as the active layers, must first be checked to achieve the aforementioned goal. There will be an equally big potential barrier causing less charge transport injection if there were space between both the different energy levels.

Either the suitable provide source with the preferred degree of work feature is required, or even the electrode areas must be chemically functionalized to transition the level of energy to the suitable place concerning the chosen electrode material, as will be covered later. To ensure the best charge carrier transport, the organic compounds in the transition region must therefore be properly arranged. Here, the appropriate molecular design improves the properties through surface treatment, and thermal austenitizing all work together to continue improving orbital overlap, which in turn results in improved conduction electrons hopping. To achieve the highest capacitance, the dielectric film should be used. This is accomplished through either would use a substance with a high electrical conductivity or cutting the width of the dielectric film. The voltage level needed to turn the gadget "on" is inversely proportional to the thickness of the dielectric layer. It is crucial to guarantee film homogeneity when depositing thin dielectric layers to reduce leakages via significantly thinner areas or faults left within.

The dielectric layer must also aid in encouraging the semiconducting layer put on top to self-organize. Lastly, to maximize the OFET effectiveness, the device measurements should be carefully chosen. For instance, shortening the channel length increases mobility while also decreasing the likelihood that charge carriers would encounter flaws or amorphous areas that will obstruct their passage (Rogers et al., 1998).

As discussed briefly above, maintaining such aspects in consideration is helpful to improve the OFET achievement. It is also helpful to be aware of all the attempts made in just this context, as discussed here. OFETs have recently been investigated for use in various display backplane applications. The better deal of the –Si devices has already been surpassed by charge transporter vibrations in the disappeared- and solution-processed components in the interim. According to a recent assessment, by the middle of 2011, there was around 40 distinct OSC with mobilities more than 1 cm^2/Vs. Unfortunately, having access to so many OSC, only a select insufficient might go to the degree of extensive addition required for show backplane arrays, maybe because of major reliability problems (Rogers and Huang, 2009).

3.4. RELEVANCE OF FULLERENES, NANOTUBES, AND GRAPHENE IN OSC DEVICES

Although significant progress has been completed in creating high-quality organic matter, flexibility has remained comparatively low, with the majority of polymer composites showing mobilities between 0.1 and 0.6 cm^2/Vs. The introduction of a preservative into the polymeric medium that produced approximately which imitated the characteristics of either of the elements involved became a way to increase mobility without replicating new technologies. The best additives for OSC-based composite material, out of the many that are accessible, were discovered to just be carbon allotropes. This first appeared using carbon fullerenes and nanotubes (CNT) using P3HT in OPV solar cells (Huang, 2009), wherein fullerene quickly emerged as the logical option owing to its superior electrical characteristics and ease of dispersion within the organic matrix in bulk organic solar cells. While CNTs were the next addition, the efficiency of solar cells was not improved by their one-dimensional (1D) architecture; instead, their use in OFETs seemed more talented (Sekitani et al., 2008).

The use of CNTs has many benefits, including very high conductance and mobility values, smaller band and also very low minimum voltages, and highly effective charge transport transit down the length of the CNT due to the very effective conjugated occurring and along the axis. Single-walled CNTs (SWCNTs) operate as both the semiconductors and conducting objects, and because of this, they should not be utilized alone since they would fault current the transistor units, unlike multi-walled carbon nanotubes (MWCNTs), where every other unit has always been conducting. For technological applications, it is thus recommended to scatter mixed CNTs in such a carbon-based matrix. MWCNTs produce bundles relatively readily due to their high surface-to-volume ratios and extended–conjugation, which can be verified by adding stabilizers with organic compounds that really can covalently attach to the outer layer. This first seems to be beneficial, but in the end, it interferes with electron delocalization, producing the creation of charge trap sites, which impairs electric transmission. Furthermore, these organic molecules disrupt the connection between CNT and the polymeric matrix, creating an extra activation energy barrier inside CNT composites. That's also valid for graphene layers, which are created by splitting open SWCNTs to create a single, thin layer. Because graphene is a two-dimensional (2D) C-allotrope, the elaborate system of–orbitals on its superficial causes an alike agglomeration to that seen in CNT bundle formations. This aggregation requires a considerable amount of power to break (Service, 1997).

3.5. HISTORICAL DEVELOPMENT PERSPECTIVES

Though research on the photocatalytic activity of anthracene crystals started in the early 20th century, the 1960s discovery of electroluminescence (EL) in crystalline materials reignited interest by revealing and examining the fundamental mechanisms underlying optical activation and charge transport transfer. The use of natural electroluminescent diodes was hampered by a few issues, despite their successful demonstration. As an illustration, issues with material stability made it challenging to maintain enough current to generate an adequate light output. Poor-quality connections and extremely high operating voltages were the results of just using substantially thicker materials, ranging in thickness from microns to millimeters. A very significant class of OSC was created as a consequence of enhanced material production and carefully managed doping of conjugated polymers over an additional 10 years. The first few uses of organic compounds began as the layer represented or photoreceptor cells in electrophotography along with the development of organic photoconductors and conducting polymers (de Boer et al., 2004; Coropceanu et al., 2007).

With both the development of two significant kinds of devices, notably solar photovoltaic cells using p- and n-type elements in a heterojunction arrangement and OFETs made from conductive polymers and oligomers, undoped OSC began gaining greater interest from scholars in the 1980s. Thus, after creating high-performance electroluminescent systems using vacuum-evaporated organically thin films and OFETs using linked polymers, the advancement was significantly amplified. OLEDs have advanced to the point where a commercial version using OLED displays is available as a result of both the continuously continuing to improve performances attained in creating better-quality components and constructing increased strategies over the past several periods (Anthony, 2006).

3.6. HIGH-MOBILITY OSC THIN FILMS

It is reasonable to reflect thin films generated from such substances as a backup alternative despite the expense of somewhat compromising the related mobilities owing to the intrinsic technical issues with the usage of monocrystalline natural matter in device manufacturing. When using thin films instead of crystals, movement is reduced by around an order of magnitude, but this seems like a fair price to pay for the extra benefits of flexible substrates on substrate surfaces that may be used in more varied applications (Aviram, 1992).

The pressure distribution determines the random orientation of the goal material in the presence of the total population of impurity atoms and goal compounds present just at the substrate all through film development, particularly at the beginning stages of both the film statement. This is how total organic films are deposited in vacuum systems. When thin films are being deposited utilizing molecular beams, bell jar depositing, and glass-wall vacuum sublimation systems, several types of pumps are employed to provide ultra-high vacuum (UHV), high vacuum (HV), and lower vacuum (LV) inside the range of 10^{-9}, 10^{-6}, and 10^{-3} torr, correspondingly. Consequently, in UHV, HV, and LV settings, the effectiveness of eliminating undesired contaminants is extremely high, moderate, and low, correspondingly (Yassar et al., 1992; Noda et al., 1999). Additionally, the carrier gas is being used to move organic compounds from the source towards the substrate, where the potting medium temperature and deposition proportion are the two factors that have a significant impact on the thin-film morphological features and the transport properties of OFETs made from of the substance so full-grown. The entire integrity of an OFET is finally ensured by ultrahigh purity predecessors and ultraclean substrate because they have a direct impact on charge transport accumulation, which is typically restricted with the first few OSC monolayers just at contact with the insulator (Noda and Shirota, 1998). Thus, the contaminants at the developing interface have an impact on the OFETs' mobility, on/off rate, and occasionally even polarity. For instance, pentacene that has been doped with iodine is a p-type material, but pentacene that has been doped with an alkaline metal has n-type behavior. The development of such devices continued in the early 1990s thanks in large part to a vapor-grown polycrystalline 6T and -di-hexyl-sexithiophene (Guay et al., 1992) film-based OFETs, which not only demonstrated that comparatively high mobilities were possible in polycrystalline natural films and also helped establish guidelines for selecting the appropriate kinds of methodologies that proved to be essential for enhancing the performance of the device. For example, thiophene oligomers are a kind of chains or rod-like molecules., π-conjugations the tight molecular stacking along at least several of the small molecular axes and the long axis of both the molecule (π-stacking) were identified as two essential requirements for attaining high transporter concentration, which held also for OFETs based on thin films containing polycrystalline pentacene that were produced by vapor (Bäuerle et al., 1995).

Developing formless pentacene films just at significantly lower temperatures of–196°C resulted in insulating films so because predominant disorders that were frozen in solid even at a low temperature prevented enough overlap between the nearest-neighbor molecule' molecular orbitals. As opposed to this, the depositions under room temperature produced highly organized films with a respectably highly mobile of 0.6 cm^2/Vs. The related transport mobility was also found to be extremely poor when a combination of amorphous with mono-crystalline stages was used to form thin films, probably because of the presence of significant defects concentration brought on by the coexistence of both the two phases.

Utilizing shadow masks, a detailed analysis of pentacene films started growing at room temp on SiO_2 reported the existence of single-crystalline islands during the early stages of growth, followed by successive layers starting to grow on top of these islands that have been smaller in size, resulting in a terrace-and-step type of morphological features. The fabrication of polycrystalline pentacene films to grain size distribution near 100 m on clean Si (001) areas please indicate with a cyclohexene layer was reported, in which the fairly low nucleation concentration of pentacene particles on such substrates besides the lack of nucleation sites just on clean Substrate surfaces were both cited as causes of the larger particle growth. Under the same depositing circumstances, SiO_2 surface nucleation frequency was 100 times greater than that of Si (001) and cyclohexene-modified Si (001) surface nucleus density. Furthermore, it was said that polycrystalline pentacene films were produced on polyimide substrates having grain sizes of 100 m utilizing a substratum temperature lower than 200°C (Nakanishi et al., 1998).

REFERENCES

1. Ahmad, S., (2014). Organic semiconductors for device applications: Current trends and future prospects. *Journal of Polymer Engineering, 34*(4), 279–338.
2. Anthony, J. E., (2006). Functionalized acenes and heteroacenes for organic electronics. *Chemical Reviews, 106*(12), 5028–5048.
3. Aviram, A., (1992). *Molecular Electronics-Science and Technology* (No. 262). Amer Inst of Physics.
4. Bao, Z., Feng, Y., Dodabalapur, A., Raju, V. R., & Lovinger, A. J., (1997). High-performance plastic transistors fabricated by printing techniques. *Chemistry of Materials, 9*(6), 1299–1301.
5. Bäuerle, P., Fischer, T., Bidlingmeier, B., Rabe, J. P., & Stabel, A., (1995). Oligothiophenes—Yet longer? Synthesis, characterization, and scanning tunneling microscopy images of homologous, isomerically pure oligo (alkylthiophene) s. *Angewandte Chemie International Edition in English, 34*(3), 303–307.
6. Bharathan, J., & Yang, Y., (1998). Polymer electroluminescent devices processed by inkjet printing: I. Polymer light-emitting logo. *Applied Physics Letters, 72*(21), 2660–2662.
7. Chang, S. C., Bharathan, J., Yang, Y., Helgeson, R., Wudl, F., Ramey, M. B., & Reynolds, J. R., (1998). Dual-color polymer light-emitting pixels processed by hybrid inkjet printing. *Applied Physics Letters, 73*(18), 2561–2563.
8. Chua, L. L., Zaumseil, J., Chang, J. F., Ou, E. C. W., Ho, P. K. H., Sirringhaus, H., & Friend, R. H., (2005). General observation of n-type field-effect behavior in organic semiconductors. *Nature, 434*(7030), 194–199.
9. Coe, S., Woo, W. K., Bawendi, M., & Bulović, V., (2002). Electroluminescence from single monolayers of nanocrystals in molecular organic devices. *Nature, 420*(6917), 800–803.
10. Coropceanu, V., Cornil, J., Da Silva, F. D. A., Olivier, Y., Silbey, R., & Brédas, J. L., (2007). Charge transport in organic semiconductors. *Chemical Reviews, 107*(4), 926–952.
11. Crone, B., Dodabalapur, A., Lin, Y. Y., Filas, R. W., Bao, Z., LaDuca, A., & Li, W., (2000). Large-scale complementary integrated circuits based on organic transistors. *Nature, 403*(6769), 521–523.

12. De Boer, R. W. I., Gershenson, M. E., Morpurgo, A. F., & Podzorov, V., (2004). Organic single-crystal field-effect transistors. *Physica Status Solidi (a), 201*(6), 1302–1331.
13. De Boer, R. W. I., Klapwijk, T. M., & Morpurgo, A. F., (2003). Field-effect transistors on tetracene single crystals. *Applied Physics Letters, 83*(21), 4345–4347.
14. Dimitrakopoulos, C. D., & Malenfant, P. R., (2002). Organic thin film transistors for large area electronics. *Advanced Materials, 14*(2), 99–117.
15. Dodabalapur, A., Katz, H. E., Torsi, L., & Haddon, R. C., (1995). Organic heterostructure field-effect transistors. *Science, 269*(5230), 1560–1562.
16. Friend, R. H., Gymer, R. W., Holmes, A. B., Burroughes, J. H., Marks, R. N., Taliani, C. D. D. C., & Salaneck, W. R., (1999). Electroluminescence in conjugated polymers. *Nature, 397*(6715), 121–128.
17. Gelinck, G. H., Huitema, H. E. A., Van, V. E., Cantatore, E., Schrijnemakers, L., Van, D. P. J. B., & De Leeuw, D. M., (2004). Flexible active-matrix displays and shift registers based on solution-processed organic transistors. *Nature Materials, 3*(2), 106–110.
18. Guay, J., Kasai, P., Diaz, A., Wu, R., Tour, J. M., & Dao, L. H., (1992). Chain-length dependence of electrochemical and electronic properties of neutral and oxidized soluble. alpha.,. alpha.-coupled thiophene oligomers. *Chemistry of Materials, 4*(5), 1097–1105.
19. Gundlach, D. J., Jackson, T. N., Schlom, D. G., & Nelson, S. F., (1999). Solvent-induced phase transition in thermally evaporated pentacene films. *Applied Physics Letters, 74*(22), 3302–3304.
20. Guo, X., Small, J. P., Klare, J. E., Wang, Y., Purewal, M. S., Tam, I. W., & Nuckolls, C., (2006). Covalently bridging gaps in single-walled carbon nanotubes with conducting molecules. *Science, 311*(5759), 356–359.
21. Hebner, T. R., & Sturm, J. C., (1998). Local tuning of organic light-emitting diode color by dye droplet application. *Applied Physics Letters, 73*(13), 1775–1777.
22. Hebner, T. R., Wu, C. C., Marcy, D., Lu, M. H., & Sturm, J. C., (1998). Ink-jet printing of doped polymers for organic light emitting devices. *Applied Physics Letters, 72*(5), 519–521.

23. Horowitz, G., (1999). Field-effect transistors based on short organic molecules. *Journal of Materials Chemistry, 9*(9), 2021–2026.

24. Kagan, C. R., Mitzi, D. B., & Dimitrakopoulos, C. D., (1999). Organic-inorganic hybrid materials as semiconducting channels in thin-film field-effect transistors. *Science, 286*(5441), 945–947.

25. Kline, R. J., & McGehee, M. D., (2003). In: Kadnikova, E. N., Liu, J., & Fréchet, J. M. J., (eds.), *Adv. Mater, 15*, 1519–1522.

26. Lang, D. V., Chi, X., Siegrist, T., Sergent, A. M., & Ramirez, A. P., (2004). Amorphouslike density of gap states in single-crystal pentacene. *Physical Review Letters, 93*(8), 086802.

27. Malliaras, G., & Friend, R., (2005). An organic electronics primer. *Physics Today, 58*(5), 53–58.

28. Mas-Torrent, M., & Rovira, C., (2008). Novel small molecules for organic field-effect transistors: Towards processability and high performance. *Chemical Society Reviews, 37*(4), 827–838.

29. Meijer, E. J., Gelinck, G. H., Van, V. E., Huisman, B. H., De Leeuw, D. M., & Klapwijk, T. M., (2003). Scaling behavior and parasitic series resistance in disordered organic field-effect transistors. *Applied Physics Letters, 82*(25), 4576–4578.

30. Murphy, A. R., & Frechet, J. M., (2007). Organic semiconducting oligomers for use in thin film transistors. *Chemical Reviews, 107*(4), 1066–1096.

31. Naber, R. C., Tanase, C., Blom, P. W., Gelinck, G. H., Marsman, A. W., Touwslager, F. J., & De Leeuw, D. M., (2005). High-performance solution-processed polymer ferroelectric field-effect transistors. *Nature Materials, 4*(3), 243–248.

32. Nakanishi, H., Sumi, N., Aso, Y., & Otsubo, T., (1998). Synthesis and properties of the longest oligothiophenes: The icosamer and heptacosamer. *The Journal of Organic Chemistry, 63*(24), 8632–8633.

33. Noda, T., & Shirota, Y., (1998). 5,5'-Bis (dimesitylboryl)-2,2'-bithiophene and 5,5"-bis (dimesitylboryl)-2, 2': 5',2"-terthiophene as a novel family of electron-transporting amorphous molecular materials. *Journal of the American Chemical Society, 120*(37), 9714–9715.

34. Noda, T., Ogawa, H., & Shirota, Y., (1999). A blue-emitting organic electroluminescent device using a novel emitting amorphous molecular material, 5,5'-bis (dimesitylboryl)-2,2'-bithiophene. *Advanced Materials, 11*(4), 283–285.

35. Popinciuc, M. J. H. T., Jonkman, H. T., & Van, W. B. J., (2006). Energy level alignment symmetry at Co/pentacene/Co interfaces. *Journal of Applied Physics, 100*(9), 093714.
36. Rogers, J. A., & Huang, Y., (2009). A curvy, stretchy future for electronics. *Proceedings of the National Academy of Sciences, 106*(27), 10875–10876.
37. Rogers, J. A., Bao, Z., & Raju, V. R., (1998). Nonphotolithographic fabrication of organic transistors with micron feature sizes. *Applied Physics Letters, 72*(21), 2716–2718.
38. Rogers, J. A., Bao, Z., Makhija, A., & Braun, P., (1999). Printing process suitable for reel-to-reel production of high-performance organic transistors and circuits. *Advanced Materials, 11*(9), 741–745.
39. Rogers, J. A., Someya, T., & Huang, Y., (2010). Materials and mechanics for stretchable electronics. *Science, 327*(5973), 1603–1607.
40. Rogers, J., & Katz, H., (1999). Printable organic and polymeric semiconducting materials and devices. *Journal of Materials Chemistry, 9*(9), 1895–1904.
41. Sekitani, T., Noguchi, Y., Zschieschang, U., Klauk, H., & Someya, T., (2008). Organic transistors manufactured using inkjet technology with subfemtoliter accuracy. *Proceedings of the National Academy of Sciences, 105*(13), 4976–4980.
42. Service, R. F., (1997). Patterning electronics on the cheap. *Science, 278*(5337), 383–384.
43. Sirringhaus, H., Brown, P. J., Friend, R. H., Nielsen, M. M., Bechgaard, K., Langeveld-Voss, B. M. W., & De Leeuw, D. M., (1999). Two-dimensional charge transport in self-organized, high-mobility conjugated polymers. *Nature, 401*(6754), 685–688.
44. Sundar, V. C., Zaumseil, J., Podzorov, V., Menard, E., Willett, R. L., Someya, T., & Rogers, J. A., (2004). Elastomeric transistor stamps: Reversible probing of charge transport in organic crystals. *Science, 303*(5664), 1644–1646.
45. Yassar, A., Delabauglise, D., Hmyene, M., Nessak, B., Horowitz, G., & Garnier, F., (1992). Synthesis and characterization of a tetra-alkylated alpha-conjugated duodecithiophene. *Advanced Materials, 4*(7, 8), 490–494.
46. Zhang, Y., Petta, J. R., Ambily, S., Shen, Y., Ralph, D. C., & Malliaras, G. G., (2003). 30 nm channel length pentacene transistors. *Advanced Materials, 15*(19), 1632–1635.

CHAPTER 4

INTRODUCTION TO OPTOELECTRONIC DEVICES

CONTENTS

4.1. INTRODUCTION

It is evident from the conversation that took place in the preceding chapters that a significant amount of effort has been done to determine the fundamental physical principles that govern electrical devices that are created using nanotubes. Such electronic devices, when they become more understood, will be able to be exploited as the essential building blocks to attain extra functionality; one instance of a developing field of study in just this respect would be that of optoelectronics using carbon nanotubes (Ruini et al., 2002; Islam et al., 2004). Moreover, the field of carbon nanotube photonics has attracted a lot of attention recently because nanotubes have several characteristics that make them great materials for optoelectronics. For instance, the existence of a wide bandgap is just an essential quality of optoelectronic substances. This bandgap makes it possible for electronic transitions to occur between both the conduction band and the valence bands even without the participation of phonons. In this regard, nanotubes stand out from other materials since they possess a straight bandgap over the whole spectrum. Therefore, for a particular nanotube, several bands may engage in straight optoelectronics activities, and these bands cover a large variety of energies. These bands can participate in these events. This should be feasible to achieve a practical way to address a wide range of wavelengths by integrating numerous nanotubes that have various bandgaps. This should be feasible via the use of numerous nanotubes. The existence of imperfections, which may be found in typical larger particles, is another issue that might arise (Léonard and Tersoff, 2002; Spataru et al., 2004). These faults can result in non-radiative recombination activities and a large decrease in the advantages of this technology. Due to the low dislocation density that they possess, nanotubes will be less susceptible to the effects of this issue. The temperature dependency of the carrier density is an additional benefit of nanotubes. This property gives nanotubes several applications. Because the concentration of states declines as it approaches the band edge in three-dimensional semiconductors, the carriers enter the channel and reach their maximum value only slightly just above the valance band. Changing the temperature causes a shift in the location of the peak, which in turn affects the characteristics of a device. Furthermore, in the case of nanotubes, the concentration of states reaches a singularity near the valence band, and this is also the location where the carriers enter the channel and reaches its highest point (Chang et al., 2004).

Just at the period that this book was being written, both theoretical and practical studies just on optical and electrical characteristics of nanotubes were only beginning to emerge. In this section, we will begin by discussing the optical characteristics of carbon nanotubes. After that, we will go on to cover three different areas of optoelectronics: photocatalytic activity, electroluminescence, as well as optoelectronics (Zhao and Mazumdar, 2004).

4.2. OPTICAL PROPERTIES

4.2.1. Excitons

The single molecule model as well as the band-to-band transition covered in the preceding section provide a good description of the optical characteristics in the majority of conventional materials. However, numerous effects predominate within optical characteristics of materials with decreased dimensions, including such polymerization or nanowires; carbon nanotubes are everywhere. Excited electrons, which are electron-hole pairings coupled by the Coulomb repulsion, are the most significant optical property hallmark of many-body events. Excitons are shown simply in Figure 4.1. According to this diagram, a positive charge hole results from the excitement of electrons from across the bandgap caused by absorbing a photon (Perebeinos et al., 2004; Zou and Zhang, 2012).

The attractive Coulomb contact between both the electron as well as the hole may result in a confined state that resembles hydrogen, with the exciton diameter separating the electron and also the hole. Similar to a hydrogen ion, the attraction potential may result in a quantization energy state; the exciton binding potential is the differential between both the free electron potential as well as the activity levels. (In actuality, there is also a repellent interaction within the electron-hole connection, which is covered in more detail below.) (Wang et al., 2005).

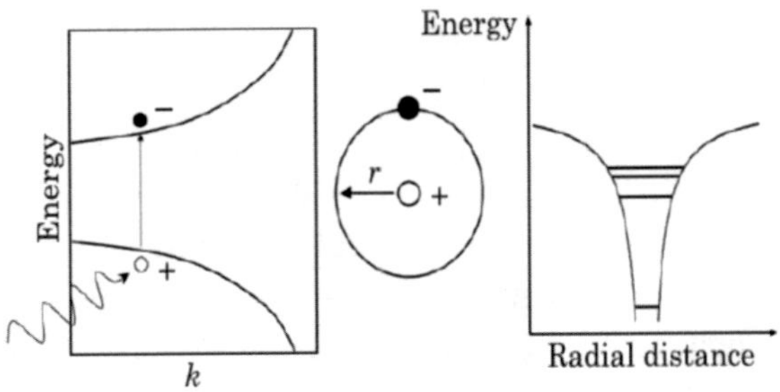

Figure 4.1. Illustration showing the production of the exciton. The simplified band structure and the excitation of one electron from the valence to the conduction band in response to photon absorption are shown in the left panel. The center panel shows that the electron may interact with the hole left behind from the Coulomb interaction, which may result in a bound state resembling hydrogen quantized energy levels in the Coulomb potential between the two particles (right panel).

Source: cambridge.org/core/books/abs/applied-nanophotonics/electrons-in-potential-wells-and-in-solids/D1A1D672320B55539DE286196D51EF47.

Examine the formula again for the lowest energy state of such a hydrogen ion to gain a straightforward demonstration of the binding affinity of excitation energy and its significance in various materials (Saito et al., 2000).

$$E_b^H = -\frac{m_0 e^4}{8h^2 r^2 \varepsilon_0^2} = -13.6eV$$

where; m_0 remains the free electron figure; ε_0 seems to be the free storage permeability; while r seems to be the Bohr radius. At least two factors contribute to the significant reduction in binding affinity in solid objects: first, the effective mass is often lower than that of the weight of something like the free electron, and secondly, screened of a Coulomb recent activity in a significantly larger dielectric permittivity. Thus, maintaining the very same excited-state radius in a straightforward model, we obtain (Perebeinos and Avouris, 2007).

$$E_b = -13.6eV \frac{m^*}{m_0}\left(\frac{\varepsilon_0}{\varepsilon}\right)^2$$

By way of an instance, for GaAs, the electron actual mass $m* = 0.067\ m_0$ and $\varepsilon = 12.85\ \varepsilon_0$ as well as the required to excite energy may be approximated to be roughly 5 meV; as a result, in conventional semiconductors, excitons often only have significance at very low temperatures. It should be noted that a key factor in the large reduction of the binding energy is the filtering of the Coulomb repulsion between the electron as well as the hole (represented either by dielectric permittivity). A quasi-1D substance, however, (π–one such example is conjugated polymer chains.) Weak electrostatic filtering results in a large increase in binding affinity. Moreover, theoretical literature has shown that the dielectric of semiconducting carbon nanotubes must be equivalent to 1, and the electrostatic interactions filtering has previously been explored in chapter concerning the characteristics of carbon nanotube p–n junctions (Stewart and Léonard, 2004).

The relevance of excitation energy in carbon nanotubes has been shown via both experimental and theoretical analysis. Just on the theoretically front, Bethe-Salpeter-based ab initio simulations of the spectra of carbon nanotubes containing conduction electrons have shown that semiconducting carbon nanotubes have a high excited-state binding affinity and even exhibit excitonic interactions in metals nanotubes (Perebeinos and Avouris, 2007). The predicted optical characteristics of an 8-nanometer carbon nanotube both with electron-hole interaction are shown in Figure 4.2 (solid line) (dashed line). The band-to-band absorptions anticipated from the single situation are indeed the peaks with the labels A, B, and C, which correlate to photons of 2.54, 2.66, and 3.7 eV, respectively. The optical absorbance spectrum is entirely changed by conduction electrons.: Every band-to-band conversion results in a string of distinct excitonic lines, denoted by A1 _, A2 _, A3 _, B1 _, B2 _, C1 _, C2 _.

A most significant finding is that now the lowest-energy excitons from the various band-to-band maxima correspond to charge carrier binding affinity of roughly 1 eV, which is substantially higher than that of conventional semiconductors (Lin et al., 2001; Freitag et al., 2003). The optical spectra of conducting polymer carbon nanotubes are thus dominated by excitons. The length of the pair of electrons through the carbon nanotube is depicted in Figure 4.2(b) and 4.2(c), where the intensity of the two-particle wave equation is displayed and the hole's location is maintained at the dot (z = 0). About 2.5 nm is discovered to be the exciton radius. Carbon nanotubes can display resonance excitons (with energy just above the solitary gap), such as the C excitons, in contrast to confined excitation energy (with energies just below the solitary gap). It is rather surprising to find a bound excited state, particularly one with strong affinity energy. It's been suggested that the metallic testing of a Coulomb repulsion, which results in an attractive delta-function possibility all along the nanostructure axis as well as an appealing delta-function possibility with a solitary bound form in one aspect, is the cause of the solitary bound exciton's existence in just this system ((Perebeinos and Avouris, 2007). Calculations show a distinct behavior for the (5,0) nanotube, where it is discovered that excitons have a little impact just on the spectral range (Marcus et al., 2006).

The electron-hole relationship also includes a repellent component, as was already discussed before in this chapter. Due to symmetries, the attraction component of interaction is muted within the (5,0) nanotube as well as the repelling term takes precedence, which prevents the emergence of bound exciton forms (Figure 4.3) (Lee, 2005).

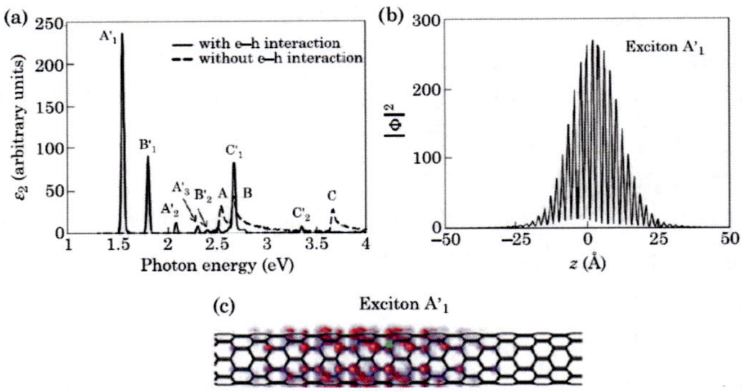

Figure 4.2. *(*a) Calculated optical absorption spectra for an 8-nanometer-long nanotube with (solid line) and without (dashed line) electron-hole interactions. (b) and (c) Shown on the nanotube and as a function of distance along the nanotube axis, respectively, are representations of the A1 exciton wavefunction. At the dot in, the hole location is determined (c) and is at z = 0 in (b).

Source: https://www.researchgate.net/figure/Absorption-spectra-for-the-carbon-nanotube-Tg-Car-oligomer-and-Tg-Car-CNT-composite_fig2_267748107.

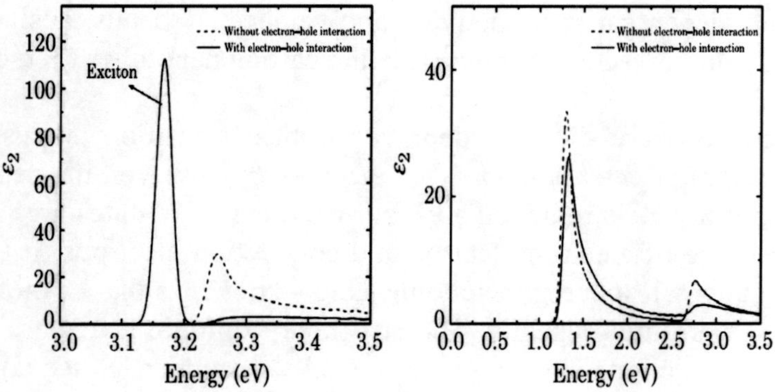

Figure 4.3. Calculated metallic carbon nanotube optical absorption spectra. The left panel depicts a single bound exciton state for a carbon nanotube with a diameter of 3, whereas the right panel depicts no bound exciton states for a nanotube with a diameter of 5,0.

Source: https://www.degruyter.com/document/doi/10.1515/nanoph-2021-0728/ html.

In contrast, semiempirical, tight-binding designs (Itkis et al., 2006) are beneficial to obtaining a physical understanding just on the contribution of excitation energy and also to research trends like the scalability of the exciton energy bands with nanotube radius. Whilst also ab initio estimations could provide thorough results for particular nanotubes, their variety of potential applications is currently restricted to the relatively small nanotube radii due to the high supercomputing demands. The Pariser-Parr-Pople (PPP) technique is a well-known example of just such a model and has been widely utilized to understand excitonic reactions in polymer chains. The Hamiltonian for this model has the following form in second quantization: (Misewich et al., 2003; Junaid et al., 2020).

$$H = -\gamma \sum_{\langle i,j \rangle,\sigma} c_{i,\sigma}^{\dagger} c_{j,\sigma} + U \sum_{i} n_{i,\uparrow} n_{i,\downarrow} + \frac{1}{2} \sum_{i,j} V_{ij} (n_i - 1)(n_{j-1})$$

$$V_{ij} = \frac{U}{k\sqrt{1 + \varsigma r_{ij}^2}}$$

where the value, using an analogy using polymerization $\varsigma = 0.6117$ is used. This semi-empirical model allowed researchers to draw several significant findings. The first is that the excited-state structure is significantly impacted by the band decadence from the single-particle picture. As shown in Figure 4.4, Double-degenerate conductivity, and valance subbands, designated as a, b and a, b, are present in semiconducting carbon nanotubes (Tersoff et al., 2005).

One can observe the four depraved optical transition processes in a single photograph. $a \rightarrow a$, $a \rightarrow b$, $b \rightarrow a$, $b \rightarrow b$; However, the degeneracy of these four layers is removed and four charge carrier states are produced within presence of electron-electron and conduction electrons. It turns out that only the greatest energy excitonic state—known as that of the "bright" exciton—is permitted visually. The last three excitonic levels are referred to as "black" excitons since they are visually forbidden. Since an exciton may rapidly relax into the visually forbidden dark exciton stages after being produced within the greatest energy level, this excitonic architecture has significant effects on the quantum yield of PL and prevents the photon from being reemitted. This is especially true since it has been discovered that the dark and brilliant excitonic phases are separated by a factor of many times kT, which prohibits the light exciton from being thermally populated (Chen et al., 2005; Marty et al., 2006).

Excitonic characteristics may be calculated using the strong binding method in a variety of nanotube settings and diameters. The predicted exciton binding potential of carbon nanotubes is shown in Figure 4.4 as a consequence of the atmosphere's dielectric constant (Pop et al., 2005).

As anticipated, the surroundings reduce the binding affinity and screening of the electrostatic attraction; the trend shows a power law connection with just an exponent independent of both the nanotube diameter. It is discovered (Figure 4.5) that there is a scaling connection between both the exciton energy bands and the carbon nanotube radius *R*, nanotube actual mass *m*, and insulator constant of the setting *ε*.

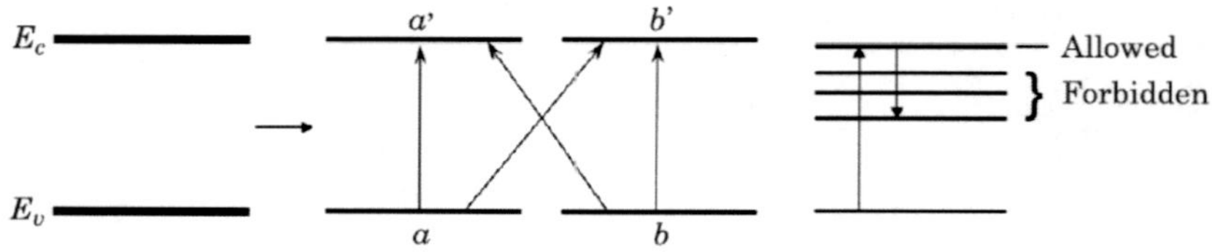

Figure 4.4. Left: A single wall carbon nanotube's doubly degenerate conduction and valence bands are shown in an example. Middle: the four optical transitions that might result from a single-particle image. Right: The four-degenerate single-particle transitions are divided in the presence of the electron-hole in-

teraction. Optically, only the condition with the maximum energy is permitted; the other three are not. To avoid photoluminescence, an electron must first be excited to its greatest energy level before relaxing into lower energy states.

Source: https://www.researchgate.net/publication/278636616_Quantum_optics_with_single-wall_carbon_nanotubes.

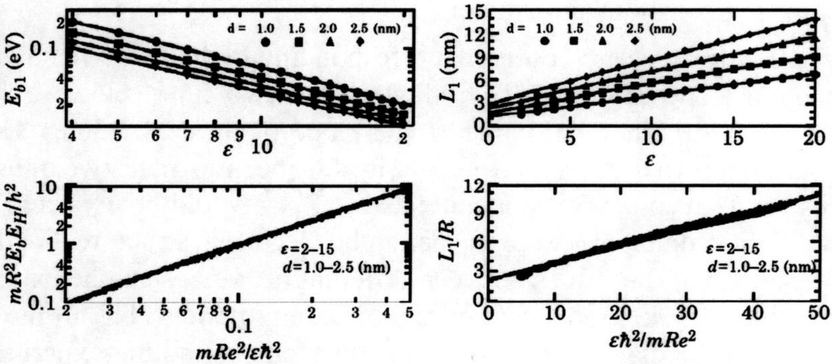

Figure 4.5. Excitons in semiconducting carbon nanotubes scale relationships. The left panels show the scaling with nanotube radius 'R' and effective mass 'm' as well as the dependency of the exciton binding energy E_{b1} on the surrounding medium's dielectric constant.

Source: https://link.springer.com/article/10.1007/s00894-020-04401-9.

Certainly, in the variety of nanotube diameters between 1 and 2.5 nm, and for insulator numbers between 2 and 15, the grading is discovered to provide a fantastic explanation of the estimated exciton binding energies.

$$mR^2 E_b = \left(\frac{mR}{\varepsilon}\right)^{1.4}$$

This scalability relationship is in line with the idea that the binding energy decreases as that of the surrounding medium's low dielectric rise, with a dependent $Eb \sim \varepsilon^{-1.4}$. Unlike bulk materials, in which the dependency is proportionate to the amount of the material, this dependency is weaker ε^{-2} (Wei et al., 2004; Phillips et al., 2006).

Excitonic actions in carbon nanotubes have recently been shown in several tests. One of the earliest investigations to offer information on the optical transition energies of particular carbon nanotubes used data from Raman scattering and PL to identify the optical transition frequencies of particular carbon nanotubes (Simmons et al., 2007). Single carbon nanotubes are disseminated in an aqueous micelle-like suspension within this research, and their optical luminosity is monitored concerning a wide variety of wavelength ranges to provide optical transition energies. The very same samples may be used for Raman spectroscopy to determine the (n,m) indices and subsequently the size of the nanotubes (Guo et al., 2005; Simmons et al., 2007). Figure 4.6 displays the observed optical transition frequencies as a function of the nanotube diameter.

Two distinct lines are observed, corresponding to the first two subbands in the optical spectrum. While each of these two lines shows a linear dependence with nanotube diameter, the slope of these two lines are not equal, and as shown in the right panel of Fig. 4.6, the ratio of the two transition energies for a given nanotube is not equal to 2, as would be expected from a simple tight-binding model of the nanotube electronic structure. While the large variations of the values between different nanotubes can be explained in part by considering more refined tight-binding models [11], in the limit of large nanotube diameters these deviations vanish, and the experimental data in Fig. 4.6 seem to converge to a ratio value clearly distinct from 2. This effect, often referred to as the "ratio problem" in carbon nanotubes, can be explained on the basis of the excitonic theory, since the first two optical transition energies are determined by the position of the two lowest energy excitons (Chang and Jean, 1999; Leonard and Stewart, 2006).

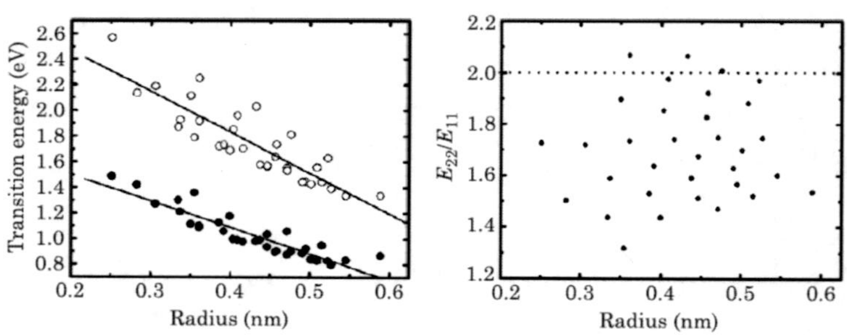

Figure 4.6. Left: photonic transition energies in semiconducting carbon nanotubes that have been determined by photoluminescence tests. Right: numbers are taken from the left panel representing the ratio of the second to the first transition energies. The ratio is projected to equal 2 in a straightforward tight-binding model, as shown by the dotted line.

Source: http://files.aiscience.org/journal/article/html/70270015.html.

Two-photon experiments provide additional scientific evidence again for the excitonic image. The optical excitations in two-photon research could be to the so-called 2p excited state or even to the continuum, as shown in Figure 4.7. The excited electron subsequently decays to the least energy exciton (1s), out of which luminescence could be seen. The crucial distinction would be that the emissions and excitement energies within the excitonic image are different from a two-photon stimulation in the band-to-band picture (Figure 4.7(b)). A contours plot of the observed emission spectra for just a sample of carbon nanotubes polymers containing matrix is shown in the right panel of Figure 4.7 as a function of the two excited states and emissions energy. The excitation source energies would've been equal to in the band-to-band model, and peaks coming from various carbon nanotubes are anticipated to lie across a line with a slope of one (solid red line in the Figure4.6). The measured peaks, with just an emissions energy substantially lower than that of the excited state, clearly diverge from such a line. The excitonic hypothesis is strongly supported by this data (Loucif-Saibi et al., 2003; Valentini et al., 2006).

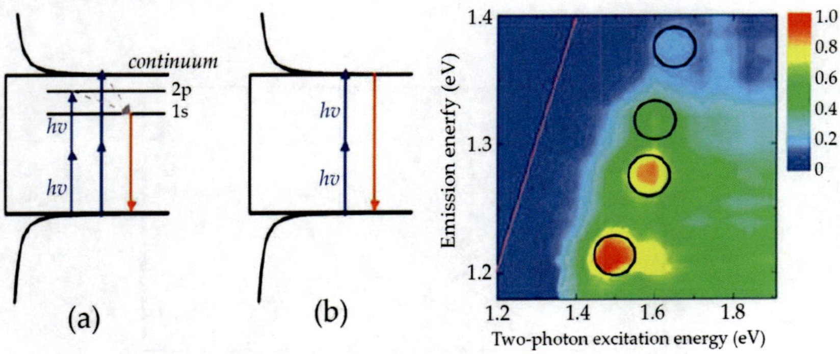

Figure 4.7. Illustration of band-to-band and exciton image (a) showing two-photon excitation (blue lines); and (b) subsequent luminescence (red lines). A color map of the observed emission intensity as a function of the two-photon excitation energy is shown in the right panel. The forecast from the band-to-band image is shown by the solid red line.

Source: https://inspirehep.net/files/4419f5874db98b6446aa3d9bf8872a36.

4.2.2. Excitons in Electric Fields

Electromagnets with in form of applied potential or band-bending as a result of doping are often used to segregate the charge carriers in optoelectronics including such photodetectors. Such electromagnetic currents might lead to an exciton's separation in the exciton model because they compete with the Coulomb attractive contact between both the electrons and holes. Therefore, it's crucial to comprehend how electric fields affect excitons to develop devices and determine whenever a subatomic pixel, band-to-band image could be acceptable. With a static homogeneous electric field present along the nanotube axis, theoretical simulations have taken this issue into account by resolving the Bethe-Salpeter equations (Cai et al., 2017; Herrick and Guo, 2021).

Figure 4.8 demonstrates how the electrical field's application affects the change in the excited-state energy bands for several semiconducting carbon nanotubes. The shift is discovered to be positive in every case, suggesting that the excited-state binding energy has increased; the actual values, though, are relatively modest, just in the range of meVs (Figure 4.9) (Choy et al., 2014).

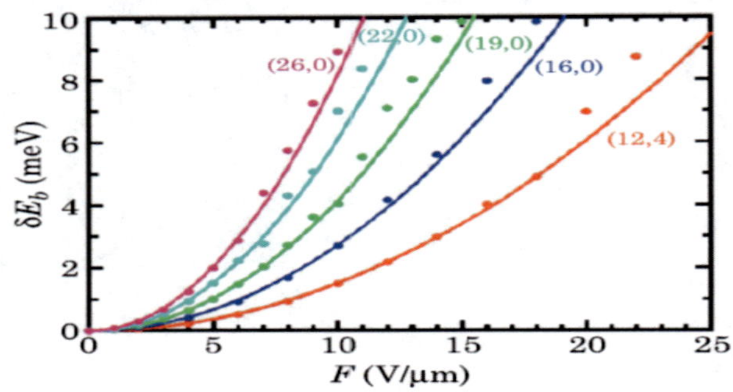

Figure 4.8. Exciton binding energy calculations for different semiconducting nanotubes as a function of the applied electric field.

Source: https://www.researchgate.net/figure/The-binding-energy-of-exciton-in-basis-on-perpendicular-electric-field-for-various-t1_fig6_260941695.

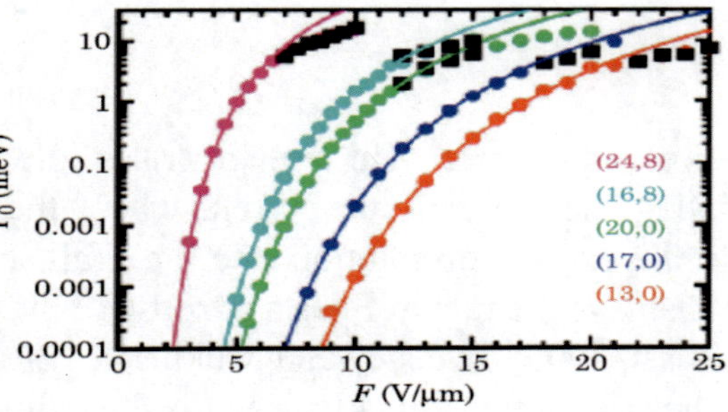

Figure 4.9. Exciton dissociation rate for various semiconducting carbon nanotubes as a function of the electric field. The nanotube indices shown from top to bottom are represented by the curves from left to right.

Source: https://aip.scitation.org/doi/10.1063/1.4963735.

An approximation of the shift's equation is:

$$\delta E_b = 4k_b \frac{(eRF)^2}{E_b}$$

with the best fit value of $k_b = 3.4$.

A bound excited state may split into unpaired electrons and a hole when a magnetic current is applied. Analysis shows that the formula can adequately explain the dissociation rate (Figure 4.10) (Hayashi, 1993; Xu et al., 2018).

$$\Gamma_0(F) = \alpha E_b \frac{F_0}{F} \exp\left(-\frac{F_0}{F}\right)$$

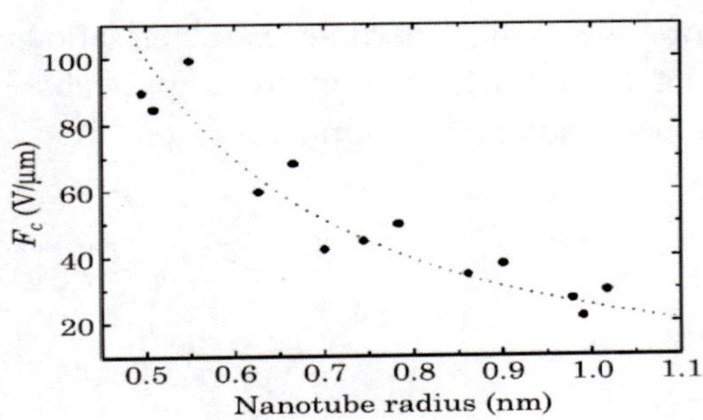

Figure 4.10. Plotting the electromagnetic current for exciton dissolution about carbon nanotube radius. The dotted line is of the form R^{-2}.

Source: https://pubs.rsc.org/en/content/chapterhtml/2021/bk9781788017824-00001?isbn=978-1-78801-782-4&sercode=bk.

where;

$$F_0 = \beta \frac{E_b^{3/2} m^{1/2}}{e\hbar}$$

As a result, even as electric field strength rises, the dissociation rate grows exponentially. The electrostatic current where the exciton binding energy is equal to the exciton dissolution rate (i.e., full separation) is $F_c \approx F_0/2$. Figure 4.11 displays the rate of F_c as a purpose of the carbon nanotube radius (Iotti and Rossi, 2005; Sweeney and Mukherjee, 2017). According to a typical pattern, the critical field quickly drops as the radius of the nanotube increases. R^{-2} trend, as shown in the illustration by the dashed lines. This performance can be found through equation and the detail that $Eb \sim R^{-1}$ and $m \sim R^{-1}$.

4.3. PHOTOCONDUCTIVITY

When a substance is lit, it may become photoconductive, which causes an electrical current to flow through it. Materials symmetric must be disrupted to produce electrons to flow in a certain direction; generally, just shining a light on the substance will not result in a net electronic flow. The p–n junction, which is at the core of several optoelectronics-like photodetectors and solar cells, maybe the simplest technology that performs in this way. The optical characteristics of carbon nanotubes were previously shown to be controlled by excitons, which may be separated in the presence of sufficiently high electromagnetic current, including those seen at a p–n junction (Chen et al., 2001; Han et al., 2018). The single-particle image is predicted to grow increasingly accurate in such a scenario. So, in this segment, we first provide a single-particle characterization of photocatalytic activity in carbon nanotubes, which ought to be applicable in circumstances where the junction has a potentially significant step.

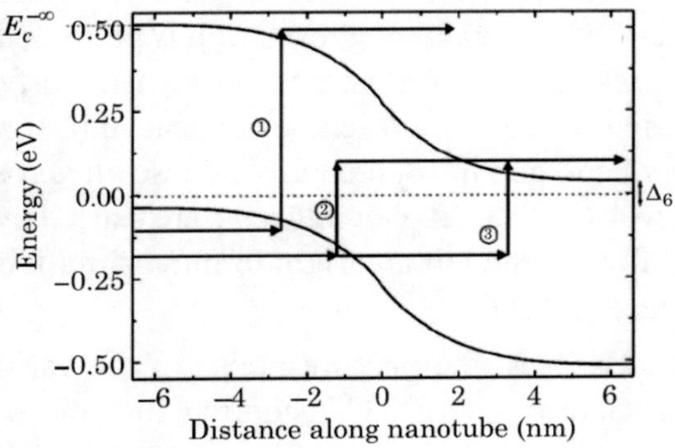

Figure 4.11. For a nanotube p–n junction and the related electronic transitions brought on by photon absorption, self-consistent band-bending calculations were made.

Source: https://www.researchgate.net/figure/Calculated-self-consistent-band-bending-for-a-CNT-p-n-junction-and-the-associated_fig39_231151769.

Even as the dynamics of excitons are better known, this concept would also be the cornerstone for understanding the mechanism of excitation energy in future (Dehghan, 2006).

Figure 4.12 shows a nanotube p–n junction's computed identity band-bending. The spatial symmetric is broken by the prospective step just at the junction. When this gadget is exposed to energy photons, $\hbar\omega$, electrons at energy E are excited to energy $E + \hbar\omega$ due to the electromagnetic current and are drawn to the electrodes (Li et al., 2006).

There in the single-particle image for a carbon nanotube p–n junction, the figure depicts the three permissible excitation mechanisms. According to Path 1, an electron may be excited from the valence band to the conduction band group by absorbing a photon with energy greater than that of the bandgap. Whenever the energy of the photon is smaller than that of the bandgap, procedures 2 and 3 demonstrate subatomic particle tunneling mechanisms (Oh et al., 2013; Liang et al., 2018).

Figure 4.12 shows the predicted photoactivity for this device at the nominal condition when 128 carbon rings are lit. The figure shows a distinct plot for the photoresponse caused by bands J = 6, 5, 7, 4, 3, and 8 (growing bandgap) (bands 11, 12, 10, 13, 14, and 9 are equivalent). A photon excites electrons in the conductive band originating first from the p-type side of the machine, moving it to the conduction band where it proceeds to the n-type terminals. When the energy of the photon surpasses the subband's bandgap, which for band 6 is 0.55 eV, such transformation is permitted. Band-to-band flow is the movement caused by electrons entering the n-type terminals in the nanotubes device inside the left lead disappear until the photon energy exceeds the energy band gap plus potential increase all across the junction., $\hbar\omega \geq 1$ eV in Figure 4.11. The net photocurrents are caused by this imbalance within the current flow to the left and right terminals in Figure 4.12 (Hoven et al., 2008; Tao et al., 2021).

As the bandgap widens, the photoresponse of the various bands tends to peak at greater photon energy. Considering that the scattering cross-sectional area shrinks as $(\hbar\omega)^{-1}$, One would anticipate that when the bandgap increases, the highest photoresponse obtained for every band will decrease. Unexpectedly, the height of the peaks in Fig. 4.12 does not follow this behavior; in particular, the response for band $J = 8$ is actually larger than the response for band $J = 6$

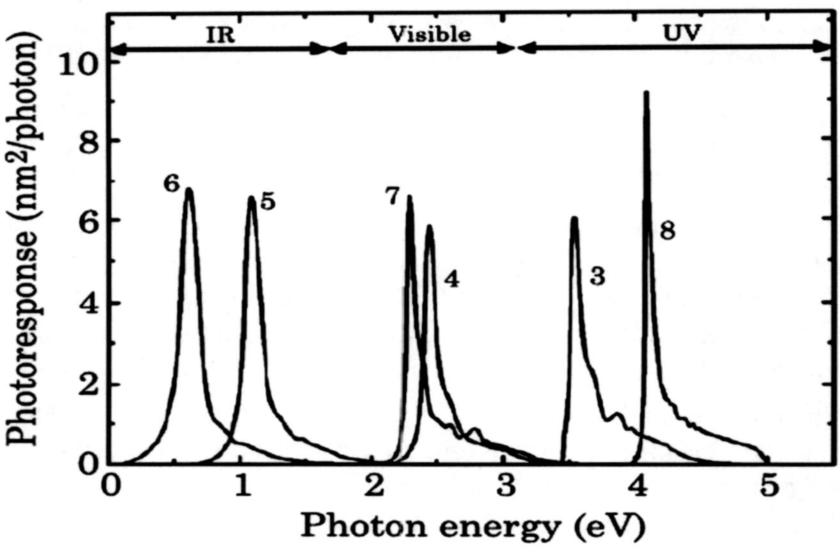

Figure 4.12. The photoresponse of a nanotube p–n junction was calculated. Each peak has an associated band of the nanotube that is identified by the angular momentum value.

Source: https://www.researchgate.net/figure/Calculated-photoresponse-of-a-CNT-p-n-junction-to-incoming-light-Each-of-the-peaks_fig40_231151769.

Specifically, the reaction for band J = 8 is bigger than that of the reaction for band J = 6 in Figure 4.12's peaks, defying this tendency. This results from the fact that the effective weights of the various bands vary. The concentration of states just at the border of the conductive band is substantially higher for J = 8 than for J = 6, due to effective mass being nearly 36 times bigger for J = 8 than for J = 6. (The significance of the state density would be covered in more detail later.) The band-to-band crossover in Path 1 of figure is shown by absorbing a photon having energy greater than the bandgap (Zhong et al., 2011; Qiao and Ma, 2020). Even while such band-to-band changes only account for a portion of the photoresponse, there is still a sizable response at an energy just below the bandgap. Photon-assisted tunneling may be credited with these contributions. Figure 4.12 depicts two potential photon-assisted tunneling routes in Paths 2 and 3. This procedure can only take place for a certain band J when $\hbar\omega \geq \Delta J$, where ΔJ is the differential (0.06 eV for the J = 6 band depicted in Figure 4.12) between the asymptotic valence band just on the n-type side and the asymptotic charge carrier edge on the p-type side. As a result, photon-assisted tunneling changes at $\hbar\omega = \Delta J$. As the energy of the photons rises above ΔJ, the photoresponse rises as more regions in the bandgap become transportable. Because tunneling probability drop with bandgap, the subatomic particle tunneling tails for bands with bigger bandgaps are less significant in comparison to a band-to-band peak (Piprek, 2016; Lin and Wang, 2018).

Thermal, visual, and ultraviolet (UV) are three different areas of electromagnetic radiation where the photoresponse of the various bands results in several sharp peaks. All of the bands inside the nanotubes have such a bandgap, which results in a response spanning a broad energy range and causes this peculiar behavior. According to the nanotube's unique electrical bandgap, which contains sets of bands isolated by comparatively large distances, this broad response has been divided into peaks clustered in several electromagnetic range areas (Figure 4.13) (Kalyanasundaram and Grätzel, 1998; Pham et al., 2019).

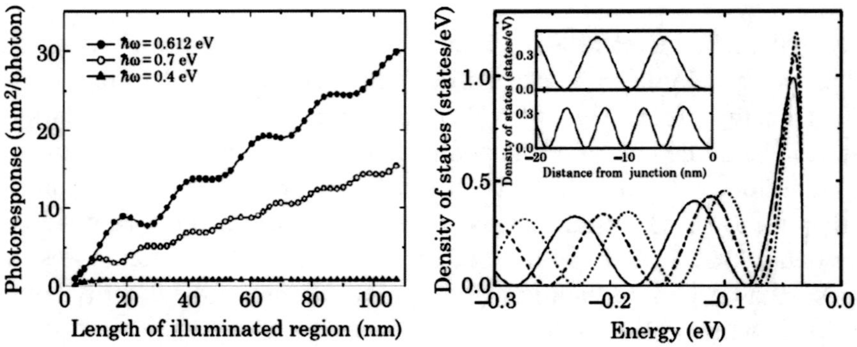

Figure 4.13. Left: depending on the length of the illuminated zone for various photon energy, the photoresponse in a carbon nanotube p–n junction. Right: energy-dependent density of states distant from the junction. For illumination lengths of 24.78, 26.88, and 28.98 nm, respectively, solid, dashed, and dotted lines are used. At an energy of 0.1 eV (0.2 eV), the top (bottom) inset displays the density of states on the even carbon rings.

Source: https://www.researchgate.net/publication/224098286_Size-dependent_light_output_spectral_shift_and_self-heating_of_400_nm_InGaN_light-emitting_diodes.

The separation of this wide response into peaks grouped in different regions of the electromagnetic spectrum is due to the particular electronic band structure of the nanotube, which has groups of bands separated by relatively large are separated from those of bands 4 and 7 (visible response) by about 0.6 eV, which are in turn separated by about 0.5 eV from the $J = 3$ and $J = 8$ conduction band edges (ultraviolet response)

The p–n junction in traditional bulk connections is normally lit via the contacts, therefore the photoresponse solely relies on the device's lengths parallel to the flow of current. However, the side-illuminated nanotube system reveals a relationship with length, as shown in Figure 4.13. Obviously, for $\hbar\omega = 0.4$ eV. The photoresponse in (subatomic particle tunneling) permeates with length as a result of the exponentially decaying of the bandgap wave equation away from of the junction. The reply for $\hbar\omega = 0.612$ eV and $\hbar\omega = 0.7$ eV demonstrates an entirely distinct pattern of behavior, fluctuating around an overall linear growth (Ha et al., 2018; Murad et al., 2020).

4.3.1. Bolometers

In addition to converting optical radiation into electron-hole pairs to produce photocurrent density, optical detecting may also be accomplished by heating a substance and causing a change in its resistance. The best example of a substance that is widely utilized is vanadium oxide in bolometer technology. Most experiments on optical detection using carbon nanotubes utilize nanotubes that are in contact with a substrate. In this case, the heat transfer to the substrate precludes the observation of a bolometric response. To circumvent this problem, a device consisting of a suspended network of carbon nanotubes has been fabricated, and has shown a strong bolometric response (Figures 4.14 and 4.15) (Perzon et al., 2006; Zhang et al., 2015).

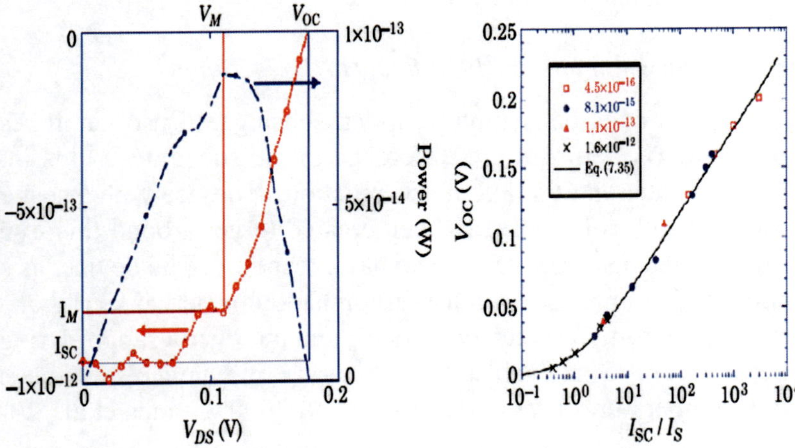

Figure 4.14. Left: For a carbon p–n junction, the current–voltage characteristics and power production. The graphic shows the short-circuit current (Isc), open-circuit voltage (Voc), and current and voltage at maximum power (IM, VM). Right: a function of light intensity for the open-circuit voltage.

Source: https://www.pveducation.org/pvcdrom/solar-cell-operation/open-circuit-voltage.

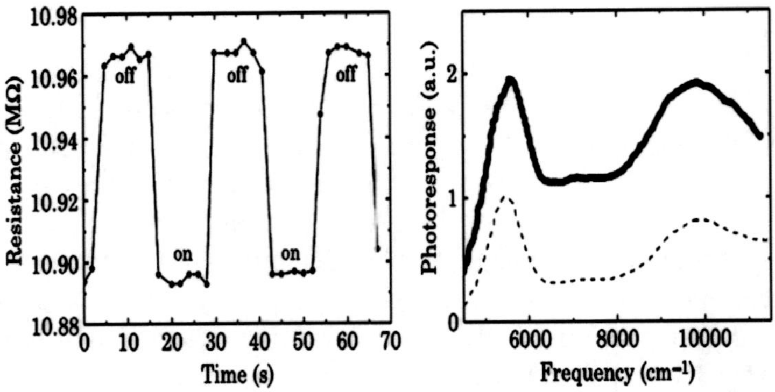

Figure 4.15. A network of carbon nanotubes strung between two electrical connections makes up the bolometer in this sketch.

Source: https://www.mdpi.com/1424-8220/19/11/2464/htm.

The majority of carbon nanotube optical sensing tests use nanotubes close to a substrate. The temperature difference to the substrate in this instance prevents the detection of a bolometric reaction. A device with a suspending system of nanotubes has already been created to get around this issue and has demonstrated a strong reflects the basic response. The device, as shown in Figure 4.15, comprises a solitary carbon nanotube networks ribbon that is 0.5 mm broad, strung between electrodes, and put over a sapphire ring's 3.5 mm aperture. A cryostat enables the measuring of a device's characteristics at various temperatures (Pauporté and Lincot, 2000; Gomaa et al., 2021).

The hanging network exhibits a significant resistance change when exposed to infrared light with anything less than microwatt power, with a transmission ratio of roughly 100. The photoresponse of the suspension network is quite well found to correlate with the absorption, which is anticipated to reflect the basic response just because more electricity is supplied to the nanotubes just at absorption coefficient peaks. In addition, similar experiments have been performed on a nanostructure system in direct contact with the floor had shown little photoresponse (Figure 4.16) (Mokkapati and Jagadish, 2009; Zhao et al., 2018).

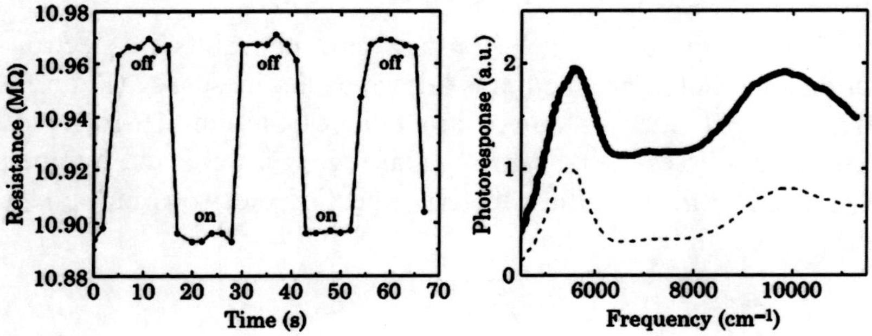

Figure 4.16. Left: As infrared photons are switched on and off, the impedance of suspended nanotube changes in the network at 50 K. Right: comparisons of an optical frequency-dependent photoresponse (solid symbols) and absorption (dotted line) of the nanotube networks.

Source: https://www.sciencedirect.com/science/article/pii/S0008622321010307.

The temperature coefficient of resistance (TCR) of a bolometric instrument must be as high as feasible. The TCR is characterized as the impedance variation with temperature:

$$TCR = \frac{TdR}{RdT}$$

This normally ranges between a few percentages in bolometer today's technology. The observed impedance of the system of nanotubes as just a function of time and temperature is shown in Figure4.16. The behavior is nonmonotonic: at cold temperatures, the resistance falls to the minimal, then rises at extremely high temperatures. The behavior at cooler temperatures is typical of semiconductors, whereas the behavior at higher temperatures is typical of metals. In the majority of bolometer applications, a negative TCR result is preferred. The bottom left panel of Figure 4.16 illustrates how the nanotubes device does this at temperatures lower than 230 K. The right panels of this figure compare, across three components with various nanotube film thickness and manufacturing circumstances, the temperature dependency of the resistivity with the temperature-dependent responsiveness. The greatest responsiveness is attained for the thinnest films (in this example, 40 nm) and is associated with the absence of the "metallic" rise in impedance with warmth. The TCR may be calculated using this information to be around 1% and 2.5% (Zeng and Liu, 2018; Bi et al., 2019).

4.4. ELECTROLUMINESCENCE (EL)

The mechanism through which constant current results in electron-hole interaction and photon production is known as luminescence (Figure 4.17). Traditional gadgets such as light emitting diodes, the simultaneous presence of electrons and holes in the same spatial region is achieved by operating a *p–n* junction in forward bias, thereby injecting electrons in the *p* region where they recombine with holes.

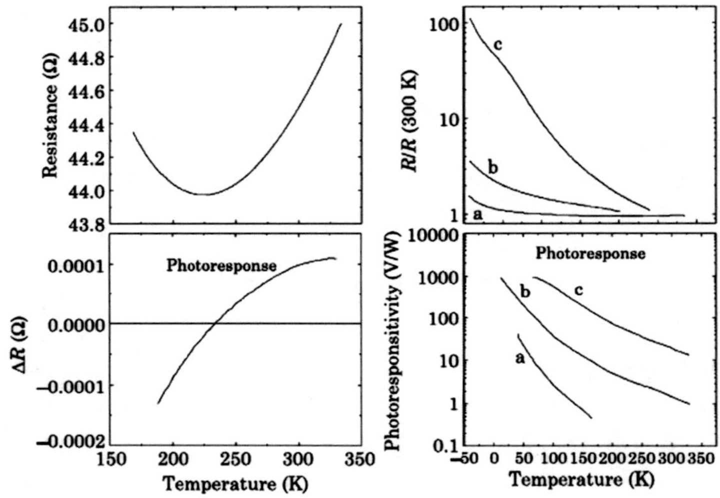

Figure 4.17. Top left: A 1-micron thick nanotube film's resistance as a function of temperature. The resistance changes with temperature when exposed to infrared light. Bottom left: Resistance for three nanotube films as a function of temperature. Upper right: (a) Purified, 1 micron thick; (b) purified, annealed in vacuum at 670 K, 100 mm thickness; (c) As-prepared, a thickness of 40 nm. The response of the nanotube films to infrared radiation is seen in the bottom right corner.

Source: https://link.springer.com/article/10.1007/s10853-021-06262-w.

By running p–n junctions during forwarding bias and introducing electrons into the p-type region, wherein they recombination with holes, light emitting diodes may accomplish the component in a mixture of electron-hole pairs in the very same spatial area. In nanotube field-effect semiconductors, one may concurrently inject electron-hole pairs into the nanotube by taking advantage of the gate manipulation of the connections and band bending. Doping is therefore not necessary in nanotube systems to see electroluminescence (EL). Figure 4.18, which depicts a schematic of ambipolar carbon nanotube transistors, serves as an illustration of the scenario. The component consists of a single semi-conductive nanotube that has Schottky barrier connections created at the source to drain by Ti (Yun, 2017). The device conductivity is managed by a strongly doped Si back gate that is isolated from the nanotubes by a 150 nm thick oxide. The device exhibits ambipolar behavior when used as Schottky barrier transistors, with roughly equal ON current flow at positively and negatively gate voltage levels (Figure 4.18(b)). However, the band-bending may be achieved thus that the electric field just at source to drain contacts has the same signs by properly selecting the gate-source as well as drain-source voltages. Whenever the drain-source voltage is higher than that of the voltage level, this occurs. With an exponential turn-on of a flow with drain current in this domain, the device effectively acts with a forward biased p–n junction (Figure 4.18(c)) (Kim et al., 2010; Liu et al., 2021). Figure 4.18(d) depicts a simulation of a band-bending all along carbon nanotube, with the selected gate-source voltage ratio, which should be equal to the difference between both the drain and source voltages.

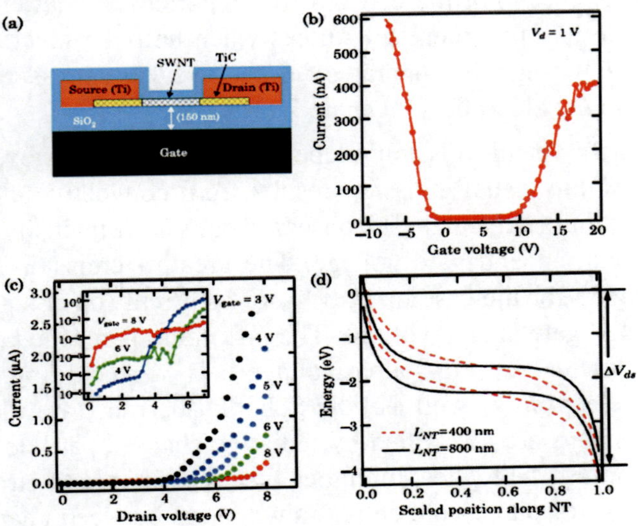

Figure 4.18. (a) A carbon nanotube electroluminescence device sketch. (b) Transfer qualities as measured. (c) The nanotube device's output properties. (d) Band bending for a nanotube device with a source-drain voltage of 4 V and a gate voltage of 2 V was calculated.

Source: https://www.mdpi.com/2079-9292/6/2/43/htm.

Electrons (holes) from either the drain (source) may tunnel through into the nanotube due to the comparatively severe band-bending around the connections. These injected electron-hole pairs could then the emission of photon recombination in the area of the nanotubes in which the bands are flat. Figure 4.18 depicts the resulting light emission (Forrest and Thompson, 2007; Dayen et al., 2021).

This picture is created by merging a photo of the carbon nanotube's optical emissions with just an infrared image of the chip's area including the nanotube device. (An infrared camera positioned on the probe stations used to test the electronic characteristics is utilized to monitor the optical emission. The picture is created by integrating the emissions over 190 seconds, and light in the 800–1,500 nm range is identified.) The yellow portions in this figure represent the metal pads and lines, whereas the blue parts in the top plane correlate to silicon dioxide. While the smaller lines serve as the source and drain terminals for the nanotube devices, the big pads, with an area of around 70 square microns, are employed to establish contact with the probe station. The emissions peak is confined to the location of the carbon nanotube, and this is a crucial aspect. Considerable evidence that the thermal photoelectric effect is not the cause of the reported thermal emission can be found in the fact that identical studies when unipolar electrons or hole currents passed through the apparatus did not reveal any measurable optical emissions (Yang et al., 2008; Wu et al., 2021).

Analyzing the visual emission's dependency on gate voltage reveals more proof that now the visual emission results from convective electron-hole interaction within the channel. The observed emissions intensity at constant drain-source current and gate voltage. The greatest emission intensity is determined again for the case of an 8 V drain current (open squares in the picture) at a 4 V gate-source voltage. The maximum is reached at a voltage level of 2 V for just a drain-source voltage of 4 V. According to a model in which the electron, as well as hole Schottky barriers, are identical and the optic emissions are controlled by radiative charged particle recombine within the channel, the highest luminous flux is seen in both circumstances whenever the gate-source voltage is midway here between channel region voltages. In reality, the actual behavior may be accurately reproduced by a model in which the fluorescence intensity is equivalent to the lowest electrons or hole current (Singh et al., 2009; Sahoo and Naik, 2022).

The visual emission is shown to be substantially polarized, which is comparable to the photocatalytic activity of single nanotubes. An infrared polarizing was placed between both the sample as well as the infrared sensor for these observations. As shown, the light emitted is noticeably stronger on one side and considerably subdued in the other direction. The spatial area where recombination takes place may be shifted by altering the gate voltage, which is especially evident when contrasted with a scanning electron microscope of the device (Kim et al., 2020). The device is run in constant-current mode throughout this sequence of photos, and the gate signal is sweeping between 30 V and 0 V while also moving the other way. No perceptible light emitted is seen from the carbon nanotube just at starting of the gate voltage movement; however, as the gate voltage decreases and becomes less negative, a clear emissions spot stands near the drain contact as well as continues to move all along the duration of the nanofibers as the gate voltage is increased toward 0 V. The emission spot vanishes as it contacts the source's electrode, and further increases in voltage level do not result in optical emission. The emissions point shifts from the source electrode to the drain electrodes as a result of a reversed gate voltage sweep. The length of the nanotube, in this instance 50 microns, is crucial to understanding this discovery. The spatial area of the recombine is substantially less than the channel width, allowing for the observation of the confined and mobile light emission (Johnston, 2013).

In terms of quality, it is possible to comprehend how this device functions by thinking about the band-bending seen in Figure 4.18. The drain current is 10 V as well as the entrance voltage is 40 V in the first picture of the series on the superior surface of Figure 4.18. In this instance (bottom left drawing), holes being inserted at it from both contacts, the electromagnetic currents at the source to drain possess opposite polarity, and also no light emitted is seen. Electron-hole pairs are concurrently injected into the nanotube channel when the gate-source voltage is progressively less negative (middle drawing), and so as a result, light is produced by recombination rate as the electromagnetic currents at the contacts ultimately have had the opposite polarity. Electromagnetic currents at both connections favor electron-hole as the gate voltage decreases toward zero, hence no light emission happens in this situation. We now go through a straightforward mathematical model that has been suggested to explain the mobility of the luminous spot (Guan et al., 2012; Yu et al., 2014).

Whenever the unipolar nanotube circuits are exposed to a strong electromagnetic current, an entirely distinct method of light emission. By dangling the nanotube more than a trench or by applying a strong source-drain bias, such electric fields may be produced. For the suspending nanotube, a SiO_2 trench between 0.4 and 15 microns wide is created, and then carbon nanotubes are generated via the chemical method and over the trenched surface. The next step is to design palladium sources and drainage electrodes on over carbon nanotubes to create channels that range in size from 4 to 80 microns (Yin et al., 2018). In Figure4.18 , an overlaying of the effect of the balance emissions and an optical picture of a device shows that, even though the gate voltage varies by many volts, the light spot remains just in contact between both the SiO_2 as well as the trench. That's in opposition to the light spot's entrance location again for an ambipolar circuit that was previously explained. Two significant distinctions from the nonsuspended system with Ti connections, in addition to a fixed location of the emitted light, are that the light emitting strength is determined to become a ratio of 1,000 higher and that the light emission happens under monopolar transport circumstances. Due to the significant electrical field produced by the discontinuities in the dielectric permittivity, it is thought that these discrepancies result from impacting excitation just at SiO_2/trench contact.

As just a function of the gate electrode, the voltage, as well as optic luminous flux, are shown. The main panel with inset show that the fluorescence intensity depends exponential just on voltage and is highest in the monopolar transportation modes of electrons or hole conducting, that is, whenever |Vg–Vth| is high. About 103,102 photon energies, or 2 to 3 magnitudes more than the non-suspended nanotube to Schottky contacts, seem to be the effectiveness of the emission that has been measured.

Whenever an electron has enough energy to produce an excited state, which further radiatively fuses, impact stimulation takes place. The electron may be driven to the required velocities inside the presence of a strong electromagnetic current, assuming that it doesn't relax as a result of phonon scattering. For the impacting stimulation process to be successful, the band-bending must take place across a length scale that is shorter than the optical phonons mean the free path, which in carbon nanotubes seems to be in the range of 10 nm. The relation may be used to predict the threshold electromagnetic current required (Sierros et al., 2009) $F\text{th} = 1.5\, Eg/l\text{op}$; for a bandgap carbon nanotube 0.5 eV and $l\text{op}$ = 10 nm, the minimum electric field is 75 V/µm. Because carbon nanotubes have a greater optical phonon random orientation than larger particles with the same bandgap, this number is roughly two orders of magnitude lower (Sierros et al., 2009).

4.4.1. Thermal Light Emission

The development of vibrational frequency optical emissions from electrical heating quasi-metallic carbon nanotubes is just an intriguing finding within nanotubes electro-optic systems (Bouhafs et al., 1998). Metallurgical nanoparticles that seem to have modest bandgaps when tested electrically are referred to just be quasi-metallic. This visual emission varies from the typical wide blackbody radiation in heated materials except it is polarized along the nanotube axis and has a spectrum with high peaks at frequencies that corresponds to the very first two band-to-band transitions. Since heat transmission to a substrate often occurs too quickly for a significant temperature to accumulate within the carbon nanotubes, such a phenomenon is more obvious in hanging carbon nanotubes (Bouhafs et al., 1998; Zhang et al., 2015).

Additionally, investigations are carried out in an argon atmosphere to stop the oxidation as well as burning of both the nanotube since the thermally stimulated emission happens at high nanotube temperatures.

For just a carbon nanotube technology that exhibits negligible gate fluctuation across a wide voltage spectrum, it displays the observed current against gate voltage. Minor gaps or nanotube flaws are thought to be the cause of the little current fluctuation. Figure 4.19(c) depicts a transmission electron microscopy of the circuit. In just this three-electrode arrangement, the nanotubes are floated above a trench inside one stream and rest on such an oxide from the other (Leung et al., 2014; Yang and Hao, 2019).

The hanging section of both the nanotube exhibits negative temperature coefficient resistance whenever the voltage is monitored as a factor of the source-drain voltage (Figure 4.19(b)), a characteristic that's also not seen in the nanotubes that are resting just on oxide substrates. The self-heating of both the carbon nanotube, which causes enhanced phonon scattering at a larger method bias, is the cause of this phenomenon (Leung et al., 2014). A hot phonon modeling approach of the negative difference conductivity yields a nanotube temperature of roughly 1,200 K, which is primarily caused by electron warming. Panels (c) and (d) of Figure 4.20 show the measured optical emissions from the three-electrode nanotube device. A few things are worth pointing out: first, the optical emission is more pronounced in the portion of the nanotube that is suspended; second, the peak that appears at about 1.6 eV can be clearly seen in the suspended nanotube but is less obvious when the nanotube is on the oxide substrate; and third, the light emission spot is discovered to be situated in the center of the channel (Jackson et al., 1997).

These findings are in line with a hypothesis in which the nanotubes are heated by joules, resulting in a high electrical temperature. According to this framework, as shown in Figure 4.19 , the high electronic average temperature causes a wide Valence distribution that significantly occupies the metallic nanotube's semiconducting subbands, enabling band-to-band electro-optic transitions to take place and result in wavelength of photons at a specific wavelength (Bioud et al., 2019).

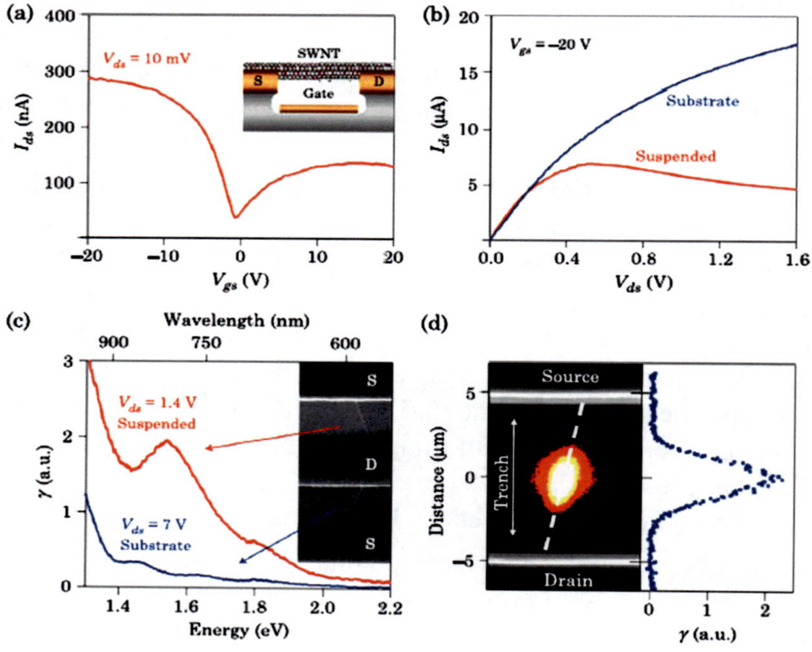

Figure 4.19. The features of quasi-metallic nanotubes in terms of electrical and optical emissions. A single carbon nanotube's power against gate supply voltages is shown in (a). The nanotube part is just on oxide substrates and the hanging section's current dependency just on drain-source voltage is shown in (b). Electron microscopy micrographs of the nanotube device and the intensity of light emitted by its suspended and unsuspended parts are shown in (c). (d) displays the measured optical emission intensity as a function of location in the channel as well as an overlay of the optical image of the device and the measured optical emission.

Source: https://pubs.acs.org/doi/10.1021/acs.accounts.7b00234.

4.5. OPTICAL DETECTION WITH FUNCTIONALIZED NANOTUBES

It's indeed possible to create photocurrent density in nanotubes utilizing monochromatic lighting at laser levels of intensity as we have indicated in the preceding sections. Even so, somewhat low illumination intensities must be converted to electronic for photodetection and switched. One method to accomplish this would be to improve the properties of carbon nanotubes to visually bioactive components using the molecular chains' responses to modify the flux in a nanotube ground triode. This method is composed of three processes in visually active molecules: (i) chromophores which change their dipole moment in the presence of light; (ii) compounds that transmit charge inside the presence of light; and (iii) compounds which, when illuminated, significantly raise scattering in nanotubes (Bioud et al., 2019). By successfully altering the gate terminal sensed either by nanotube, the poisoning of the nanotubes or by causing the current to decrease owing to greater scatters, these methods may be utilized to regulate the flow in ground transistors.

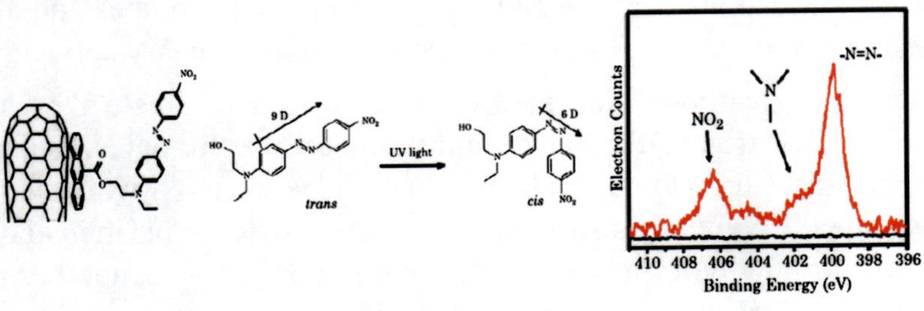

Figure 4.20. Solitary nanotubes bifunctional using disperse red 1 (DR1) through an anthraquinone linker are seen in the drawing. The equilibrium trans configuration of DR1 isomerizes to the metastable cis configuration when exposed to UV radiation. The result is a shift in the molecule dipolar from 9 to 6 D. Right: Photoemission spectra of the N(1s) core level.

Source: https://www.sciencedirect.com/topics/chemistry/disperse-red-1.

4.5.1. Modulation of Molecular Dipole Moment

Figure 4.20 shows a drawing of a carbon nanotube bifunctional with the azobenzene monomer disperse red 1 (DR1), which is used to give acrylics their red hue. Under UV light, it is known that this molecule isomerizes, causing a significant shift inside the dipole moment. According to experiments, the DR1 carbonyl group is in the trans conformational position (Figure 4.20), where there is a large orbital overlapping between both the phenyl group, when in balance (Diring et al., 2009).

The trans structure possesses a sizeable, 9 D magnetic dipole due to this conjugated as well as the existence of the highly electronegative nitro end group. The chromophore global warming to the cis configuration (Figure 4.20) when exposed to UV light, in which the orbital overlaps are drastically decreased and a lower, 6 D polarizability results. The cis isoform will relax to some less substituent restricted and so more stable trans conformation when left under ambient conditions. The DR1 chromophore is noncovalently attached to nanotubes using an anthraquinone tether to improve its properties. Even though such chromophores may be covalently or noncovalently connected to the sidewalls of nanotubes, a non-covalent interactions connection is preferable since it only slightly disturbs the electronic properties of the nanotubes. Pyrene and anthracene, two polycyclic aromatics, have already been demonstrated to be physisorbed onto other nanomaterials by forming π–π relatively negligible charge transfer and binds to the sidewalls of the nanotube. Because most π–π Noncovalent connection provides the possibility of reversibly functionalizing nanotubes and allows for the simple removal of attached compounds using ordinary solvents.

Dicyclohexylcarbodiimide methyl ester is used to create the sample containing DR1 from DR1 and 9 anthracenecarboxylic acids. Silica gel electrophoresis is used to purify the raw material, and 1H-NMR is then used to determine the company's structure. Anthracene-DR1 is purified and then dissolved in dimethylformamide before being applied to nanotubes. Also included is purified anthracenecarboxylic acid to be used as a non-functional control. By using chemical vapor deposition (CVD), lithography, and a 500 nm thermally oxide, the single nanotube devices are made on heavily doped silicon chips. Iron, as well as molybdenum nanoparticles catalysis, are used to generate the nanotubes at 900°C utilizing methane feedstock and hydrogen coflow. The typical nanotube length in these CVD circumstances is 1.6 nm. Following growth, particular nanotubes are located using electrons and transmission electron microscopes, and connections are formed using electron beam lithography, physical vapor deposition, as well as a 400°C formation gas anneal (4:1 Ar/H2). After annealed, the chip is coated with a droplet of the anthracene-DR1 (also known as anthracenecarboxylic acid) solutions, as well as the samples are then rinsed to get rid of the unintended chromophores. Indicated by the N(1s) to C(1s) intensity ratio, X-ray photoelectron spectroscopy (XPS, Figure 4.20) studies show that now the chromophore persists after cleaning and yields coverage of 1–2 molecules per 100 nanotubes carbon atoms. Isolated single-wall nanotube transistors' I-Vg properties afterwards when chromophore attachment reveal negligible variations within power dissipation, demonstrating that anthracene-DR1 molecules scatter electric charge very little and have no adverse impacts on the productivity of the device. Because of the tiny system: electrostatic interaction from anthracene, devices may show threshold voltage variations. The n-channel does not open up when the gate voltage is scanned to high positive values, proving that the semiconductors are unipolar. A portable UV lamp having 254 as well as 365 nm wavelengths and moderate intensity of 100 W/cm^2 is used to accomplish UV-induced switching. To completely remove the possibility of ozone degradation from UV stimulation of the ambient oxygen, electrical assessments were carried out in a nitrogen-purged chamber (Deng et al., 2015).

A sequence of I-Vg properties is obtained when varying the gate voltage while monitoring the draining current flowing through a nanotube transistor. The transistor exhibits p-type behavior before UV irradiation and has a voltage level of roughly 1 V. The electrical properties are moved to the right, in this example by 0.7 V, whenever the chromophore occurs when an object to the cis configuration, and it is independent of a UV light's frequency. Many nanotube semiconductors had their sub-threshold shift evaluated under lighting; all of them had positive threshold changes ranging from 0.6–1.2 V. A modification in the nearby electrostatic surroundings or a start charging process might be the cause of this shift. Control studies demonstrate no minimum voltage change, suggesting that there is no photoinduced transference from the anthracene tether. Electron transfer from the chromophore to a nanotube is prevented by the alkane spacers and anthraquinone tether between the chromophore as well as the nanotube. Electron transfer could be completely ruled out since estimation of the quantity of charge transport required for a 1 V threshold voltage change yields 0.07 e/molecule. It is suggested that now the DR1 electrode potential change functions as a modest local negative voltage level because the UV photoisomerization and contemporaneous electrode potential change are well-recognized processes (Figure 4.21). The vast transistor threshold voltage shift is due to the relatively short spacer group used to separate the chromophore from the nanotube (1 nm for anthracene-DR1).

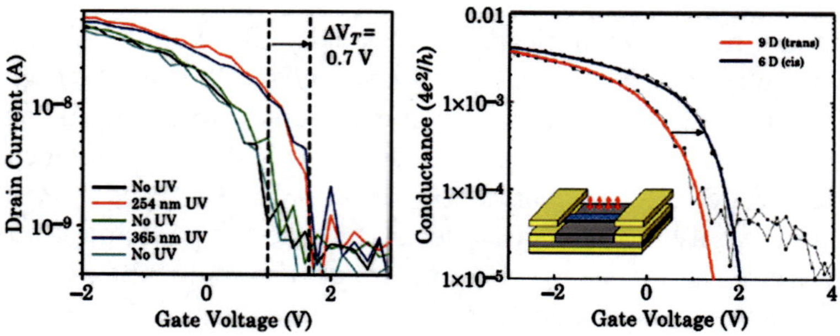

Figure 4.21. Left: Under UV light, the threshold voltage changes in transistor properties. For both 254 and 365 nm light, the threshold voltage has a 0.7 V shift to the right and is completely reversible. Comparing experimental results with device models of the nanotube conductivity for both chromophore isomers (solid lines), on the right (gray symbols). When the chromophore switches to the 6 D cis isomer, the transistor characteristic for the 9 D trans isomer (red curve, left) shifts toward positive gate voltages (blue curve, right). The results demonstrate a 700-mV change in the threshold voltage to the right. The red arrows in the inset show the molecular dipoles in the nanotube transistor sketch that was used in the simulations.

Source: https://www.researchgate.net/publication/6448333_Optically_Modulated_Conduction_in_Chromophore-Functionalized_Single-Wall_Carbon_Nanotubes.

The comparatively short spacers group utilized to isolate the chromophore from the nanotube is to blame for the change in transistors threshold voltage (1 nm for anthracene-DR1).

4.5.2. Charge Transfer

Channel permeability may be affected by methods that result in charge transport to a channel since field-effect semiconductors are devices that regulate the charges in the channel. Due to the channel's nanometer-sized dimensions, this effect may be very potent in carbon nanotubes. By optically stimulating charge transfer here between nanotubes and compounds on the nanotube surfaces, this phenomenon has been used to regulate the permeability of carbon nanotube ground transistors.

Porphyrin, a crucial component of photosynthetic, is one specific compound that is very well known to experience photoinduced transfer. Porphyrin-functionalized nanomaterials exhibit charge separation states with comparatively extended lifetimes, according to experimental research.

Exploiting chemical vapor to manufacture single-walled carbon nanotubes on either a silicon wafer with 500 nm of the gate oxide or carbon nanotube ground transistor were created to investigate the feasibility of using these photogenerated charge transfers for photodetection and switching. After patterning Pd/Cr electrodes onto the networks of nanotubes to create field-effect semiconductors with 500 m and 1,000 m channel lengths and widths, the growing method produced networks of nanotubes with a concentration of 1.6 nanotubes/m². Because the length of both the channel is significantly larger than the size of the nanotubes, primary active transport by percolation. Additionally, because only one-third of the nanostructures are metal, there are relatively few direct metallic pathways throughout the channel, assuring transistors behavior due to a semi-conductive nature.

The circuit resistance is controlled by the tube, not the connections, by lengthening the short channel effects through several 100 microns. Only the central part of the devices was accessible to the porphyrin compounds as the nanotube networks were polymerized with zinc(II) metalloporphyrin (see a drawing of molecules and devices in Figure 4.22). Given the length of the pipe and the lack of synthesis process close to the contacts, it is likely that now the contacts are not the cause of any changes in device impedance brought on by light exposure.

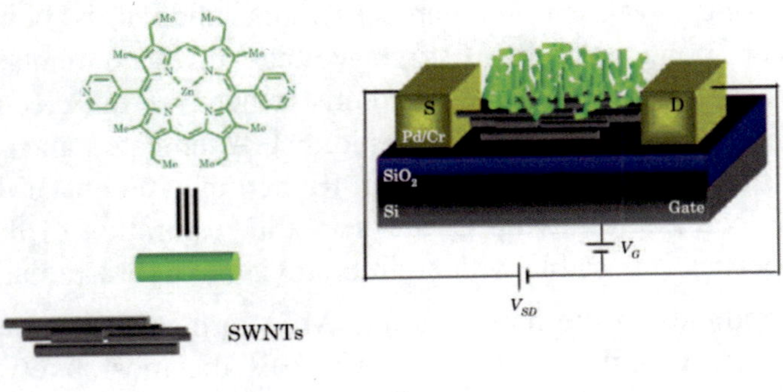

Figure 4.22. On the right is a sketch of the zinc(II) metalloporphyrin that was utilized to functionalize a carbon nanotube field-effect transistor. The transistor channel is comprised of a network of carbon nanotubes and is several microns long.

Figure 4.22 demonstrates the field-effect transistors made of carbon nanotubes' determined transfer properties during the porphyrin was deposited. Though this may be because there is an electric potential above the reported values, the bare nanotubes device exhibits a rather mild gate dependency. The gadget exhibits distinct ON and OFF modes with a threshold voltage of roughly 4 V after porphyrin functionalization. However, when the porphyrin is attached, the current's strength is reduced by almost an order of magnitude. In comparison, the non-covalent interactions bonding with the quinone linker had no impact on the amplitude of the current in the DR1 synthesized nanotubes reported in the preceding section. The compounds on the nanotube surfaces may cause an increase in conduction electrons, and/or changes in conductance at the nanotube-nanotube crossover sites may be the reason for the current drop. A voltage level in the gates voltage window indicates that the hole density in nanotubes is decreasing; this might be caused by the transfer of electrons first from porphyrin or by the expulsion of oxygen out from the surfaces of the nanomaterials. The concentration of present on the surface of both the nanotubes is discovered to affect how the transmission properties change first from porphyrin-covered behavior to the bare behavior, with both the curve of Figure 4.22 correlating to the saturation behavior.

4.5.3. Scattering

The functionalization of carbon nanotubes containing spiropyran, a chemical that changes into a charge-separated state when exposed to UV light, was shown to cause photoswitching of carbon nanotube ground transistors. Pyrene and alkane families were used to connect these substances to the surfaces of single-wall carbon nanostructures since it is believed that they connect noncovalently. Current-voltage measurements revealed that perhaps the voltage gain is not altered whenever the electronics are illuminated, in contrast to the sub-threshold shift indicated above with the dipole modifications and charge transfer. Rather, the amount of the current is shown to decrease following UV rays. A full restoration of the original transfer properties is possible with additional visible light irradiation.

The compounds in the merocyanine (MC) form of the compounds in the nanotube have higher dispersion, which is the most likely cause of the decrease in current. There might be several reasons for this behavior, but they all seem to hinge on the idea that the chemicals are not closely compacted just on nanotube surfaces. For instance, it is feasible that the distance between the compounds exceeds the decay duration of the electrical potential produced by the MC's polarization; in just this case, localized perspective perturbations or a general shift of the electrical potential are anticipated. If the conduction mechanism between both the compounds and indeed the nanotubes is confined, a similar circumstance may occur (Sun et al., 2005).

REFERENCES

1. Bi, Y. G., Liu, Y. F., Zhang, X. L., Yin, D., Wang, W. Q., Feng, J., & Sun, H. B., (2019). Ultrathin metal films as the transparent electrode in ITO-free organic optoelectronic devices. *Advanced Optical Materials, 7*(6), 1800778.
2. Bioud, Y. A., Boucherif, A., Myronov, M., Soltani, A., Patriarche, G., Braidy, N., & Arès, R., (2019). Uprooting defects to enable high-performance III–V optoelectronic devices on silicon. *Nature Communications, 10*(1), 1–12.
3. Bouhafs, D., Moussi, A., Chikouche, A., & Ruiz, J. M., (1998). Design and simulation of antireflection coating systems for optoelectronic devices: Application to silicon solar cells. *Solar Energy Materials and Solar Cells, 52*(1, 2), 79–93.
4. Cai, Y., Qin, A., & Tang, B. Z., (2017). Siloles in optoelectronic devices. *Journal of Materials Chemistry C, 5*(30), 7375–7389.
5. Cao, W., Li, J., Chen, H., & Xue, J., (2014). Transparent electrodes for organic optoelectronic devices: A review. *Journal of Photonics for Energy, 4*(1), 040990.
6. Chang, C. R., & Jean, J. H., (1999). Crystallization kinetics and mechanism of low-dielectric, low-temperature, cofirable CaO-B_2O_3-SiO_2 glass-ceramics. *Journal of the American Ceramic Society, 82*(7), 1725–1732.
7. Chang, E., Bussi, G., Ruini, A., & Molinari, E., (2004). Excitons in carbon nanotubes: An ab initio symmetry-based approach. *Physical Review Letters, 92*(19), 196401.
8. Chen, J., Perebeinos, V., Freitag, M., Tsang, J., Fu, Q., Liu, J., & Avouris, P., (2005). Bright infrared emission from electrically induced excitons in carbon nanotubes. *Science, 310*(5751), 1171–1174.
9. Chen, Z., Cotterell, B., Wang, W., Guenther, E., & Chua, S. J., (2001). A mechanical assessment of flexible optoelectronic devices. *Thin Solid Films, 394*(1, 2), 201–205.
10. Choy, W. C., Chan, W. K., & Yuan, Y., (2014). Recent advances in transition metal complexes and light-management engineering in organic optoelectronic devices. *Advanced Materials, 26*(31), 5368–5399.
11. Coleman, J. J., (1997). Metalorganic chemical vapor deposition for optoelectronic devices. *Proceedings of the IEEE, 85*(11), 1715–1729.

12. Dayen, J. F., Konstantinov, N., Palluel, M., Daro, N., Kundys, B., Soliman, M., & Doudin, B., (2021). Room temperature optoelectronic devices operating with spin crossover nanoparticles. *Materials Horizons, 8*(8), 2310–2315.

13. Dehghan, M., (2006). Finite difference procedures for solving a problem arising in modeling and design of certain optoelectronic devices. *Mathematics and Computers in Simulation, 71*(1), 16–30.

14. Deng, H., Dong, D., Qiao, K., Bu, L., Li, B., Yang, D., & Song, H., (2015). Growth, patterning and alignment of organolead iodide perovskite nanowires for optoelectronic devices. *Nanoscale, 7*(9), 4163–4170.

15. Diring, S., Camerel, F., Donnio, B., Dintzer, T., Toffanin, S., Capelli, R., & Ziessel, R., (2009). Luminescent ethynyl− pyrene liquid crystals and gels for optoelectronic devices. *Journal of the American Chemical Society, 131*(50), 18177–18185.

16. Forrest, S. R., & Thompson, M. E., (2007). Introduction: Organic electronics and optoelectronics. *Chemical Reviews, 107*(4), 923–925.

17. Freitag, M., Chen, J., Tersoff, J., Tsang, J. C., Fu, Q., Liu, J., & Avouris, P., (2004). Mobile ambipolar domain in carbon-nanotube infrared emitters. *Physical Review Letters, 93*(7), 076803.

18. Freitag, M., Martin, Y., Misewich, J. A., Martel, R., & Avouris, P., (2003). Photoconductivity of single carbon nanotubes. *Nano Letters, 3*(8), 1067–1071.

19. Gaul, D. A., & Rees, Jr. W. S., (2000). True blue inorganic optoelectronic devices. *Advanced Materials, 12*(13), 935–946.

20. Gomaa, H. M., Yahia, I. S., Makram, B. M. A., El-Dosokey, A. H., & Elkatlawy, S. M., (2021). Optical and structural studies of some zinc calcium borate glasses for optoelectronic device applications. *Journal of Materials Science: Materials in Electronics, 32*(7), 9392–9399.

21. Guan, X., Zhang, K., Huang, F., Bazan, G. C., & Cao, Y., (2012). Amino N-oxide functionalized conjugated polymers and their amino-functionalized precursors: New cathode interlayers for high-performance optoelectronic devices. *Advanced Functional Materials, 22*(13), 2846–2854.

22. Guldi, D. M., Taieb, H., Rahman, G. A., Tagmatarchis, N., & Prato, M., (2005). Novel photoactive single-walled carbon nanotube–porphyrin

polymer wraps: Efficient and long-lived intracomplex charge separation. *Advanced Materials, 17*(7), 871–875.

23. Guo, D., Guo, Q., Chen, Z., Wu, Z., Li, P., & Tang, W. J. M. T. P., (2019). Review of Ga_2O_3-based optoelectronic devices. *Materials Today Physics, 11*, 100157.
24. Guo, X., Huang, L., O'Brien, S., Kim, P., & Nuckolls, C., (2005). Directing and sensing changes in molecular conformation on individual carbon nanotube field effect transistors. *Journal of the American Chemical Society, 127*(43), 15045–15047.
25. Ha, D., Fang, Z., & Zhitenev, N. B., (2018). Paper in electronic and optoelectronic devices. *Advanced Electronic Materials, 4*(5), 1700593.
26. Han, J., Guo, S., Lu, H., Liu, S., Zhao, Q., & Huang, W., (2018). Recent progress on circularly polarized luminescent materials for organic optoelectronic devices. *Advanced Optical Materials, 6*(17), 1800538.
27. Hayashi, I. H. I., (1993). Optoelectronic devices and material technologies for photo-electronic integrated systems. *Japanese Journal of Applied Physics, 32*(1S), 266.
28. Hecht, D. S., Ramirez, R. J., Briman, M., Artukovic, E., Chichak, K. S., Stoddart, J. F., & Grüner, G., (2006). Bioinspired detection of light using a porphyrin-sensitized single-wall nanotube field effect transistor. *Nano Letters, 6*(9), 2031–2036.
29. Herrick, R. W., & Guo, Q., (2021). Introduction to optoelectronic devices. In: *Reliability of Semiconductor Lasers and Optoelectronic Devices* (pp. 1–45). Woodhead Publishing.
30. Hoven, C. V., Garcia, A., Bazan, G. C., & Nguyen, T. Q., (2008). Recent applications of conjugated polyelectrolytes in optoelectronic devices. *Advanced Materials, 20*(20), 3793–3810.
31. Hu, Z., Huang, F., & Cao, Y., (2017). Layer-by-layer assembly of multilayer thin films for organic optoelectronic devices. *Small Methods, 1*(12), 1700264.
32. Iotti, R. C., & Rossi, F., (2005). Microscopic theory of semiconductor-based optoelectronic devices. *Reports on Progress in Physics, 68*(11), 2533.
33. Islam, M. F., Milkie, D. E., Kane, C. L., Yodh, A. G., & Kikkawa, J. M., (2004). Direct measurement of the polarized optical absorption cross section of single-wall carbon nanotubes. *Physical Review Letters, 93*(3), 037404.

34. Itkis, M. E., Borondics, F., Yu, A., & Haddon, R. C., (2006). Bolometric infrared photoresponse of suspended single-walled carbon nanotube films. *Science, 312*(5772), 413–416.
35. Jackson, A. W., Pinsukanjana, P. R., Gossard, A. C., & Coldren, L. A., (1997). In situ monitoring and control for MBE growth of optoelectronic devices. *IEEE Journal of Selected Topics in Quantum Electronics, 3*(3), 836–844.
36. Johnston, A. H., (2013). Radiation effects in optoelectronic devices. *IEEE Transactions on Nuclear Science, 60*(3), 2054–2073.
37. Junaid, M., Md Khir, M. H., Witjaksono, G., Ullah, Z., Tansu, N., Saheed, M. S. M., & Siddiqui, M. A., (2020). A review on graphene-based light emitting functional devices. *Molecules, 25*(18), 4217.
38. Kalyanasundaram, K., & Grätzel, M., (1998). Applications of functionalized transition metal complexes in photonic and optoelectronic devices. *Coordination Chemistry Reviews, 177*(1), 347–414.
39. Kim, J. T., Lee, J., Jang, S., Yu, Z., Park, J. H., Jung, E. D., & Lee, B. R., (2020). Solution processable small molecules as efficient electron transport layers in organic optoelectronic devices. *Journal of Materials Chemistry A, 8*(27), 13501–13508.
40. Kim, S. H., Park, S. H., & Lee, K., (2010). Efficiency enhancement in polymer optoelectronic devices by introducing titanium sub-oxide layer. *Current Applied Physics, 10*(3), S528–S531.
41. Lan, C., Shi, Z., Cao, R., Li, C., & Zhang, H., (2020). 2D materials beyond graphene toward Si integrated infrared optoelectronic devices. *Nanoscale, 12*(22), 11784–11807.
42. Larrimore, L., Nad, S., Zhou, X., Abruna, H., & McEuen, P. L., (2006). Probing electrostatic potentials in solution with carbon nanotube transistors. *Nano Letters, 6*(7), 1329–1333.
43. Lee, J. U., (2005). Photovoltaic effect in ideal carbon nanotube diodes. *Applied Physics Letters, 87*(7), 073101.
44. Lee, J. U., (2007). Band-gap renormalization in carbon nanotubes: Origin of the ideal diode behavior in carbon nanotube p– n structures. *Physical Review B, 75*(7), 075409.
45. Leonard, F., & Stewart, D. A., (2006). Properties of short channel ballistic carbon nanotube transistors with ohmic contacts. *Nanotechnology, 17*(18), 4699.

46. Léonard, F., & Tersoff, J., (2002). Dielectric response of semiconducting carbon nanotubes. *Applied Physics Letters, 81*(25), 4835–4837.

47. Leung, S. F., Zhang, Q., Xiu, F., Yu, D., Ho, J. C., Li, D., & Fan, Z., (2014). Light management with nanostructures for optoelectronic devices. *The Journal of Physical Chemistry Letters, 5*(8), 1479–1495.

48. Li, Y., Qian, F., Xiang, J., & Lieber, C. M., (2006). Nanowire electronic and optoelectronic devices. *Materials Today, 9*(10), 18–27.

49. Liang, F. X., Gao, Y., Xie, C., Tong, X. W., Li, Z. J., & Luo, L. B., (2018). Recent advances in the fabrication of graphene–ZnO heterojunctions for optoelectronic device applications. *Journal of Materials Chemistry C, 6*(15), 3815–3833.

50. Lim, G. E., Ha, Y. E., Jo, M. Y., Park, J., Kang, Y. C., & Kim, J. H., (2013). Nonconjugated anionic polyelectrolyte as an interfacial layer for the organic optoelectronic devices. *ACS Applied Materials & Interfaces, 5*(14), 6508–6513.

51. Lin, S. Y., Tsai, Y. R., & Lee, S. C., (2001). High-performance InAs/GaAs quantum-dot infrared photodetectors with a single-sided Al 0.3 Ga 0.7 As blocking layer. *Applied Physics Letters, 78*(18), 2784–2786.

52. Lin, T., & Wang, J., (2018). Applications of graphdiyne on optoelectronic devices. *ACS Applied Materials & Interfaces, 11*(3), 2638–2646.

53. Litvin, A. P., Zhang, X., Ushakova, E. V., & Rogach, A. L., (2021). Carbon nanoparticles as versatile auxiliary components of perovskite-based optoelectronic devices. *Advanced Functional Materials, 31*(18), 2010768.

54. Liu, Z., Zou, Y., Ji, C., Chen, X., Hou, G., Zhang, C., & Zhang, X., (2021). Broad-spectrum ultrathin-metal-based oxide/metal/oxide transparent conductive films for optoelectronic devices. *ACS Applied Materials & Interfaces, 13*(49), 58539–58551.

55. Loucif-Saibi, R., Nakatani, K., Delaire, J. A., Dumont, M., & Sekkat, Z., (1993). Photoisomerization and second harmonic generation in disperse red one-doped and-functionalized poly (methyl methacrylate) films. *Chemistry of Materials, 5*(2), 229–236.

56. Luo, D., Li, X., Dumont, A., Yu, H., & Lu, Z. H., (2021). Recent progress on perovskite surfaces and interfaces in optoelectronic devices. *Advanced Materials, 33*(30), 2006004.

57. Mann, D., Kato, Y. K., Kinkhabwala, A., Pop, E., Cao, J., Wang, X., & Dai, H., (2007). Electrically driven thermal light emission from

individual single-walled carbon nanotubes. *Nature Nanotechnology, 2*(1), 33–38.

58. Marcus, M. S., Simmons, J. M., Castellini, O. M., Hamers, R. J., & Eriksson, M. A., (2006). Photogating carbon nanotube transistors. *Journal of Applied Physics, 100*(8), 084306.
59. Marty, L., Adam, E., Albert, L., Doyon, R., Menard, D., & Martel, R., (2006). Exciton formation and annihilation during 1D impact excitation of carbon nanotubes. *Physical Review Letters, 96*(13), 136803.
60. Misewich, J. A., Martel, R., Avouris, P., Tsang, J. C., Heinze, S., & Tersoff, J., (2003). Electrically induced optical emission from a carbon nanotube FET. *Science, 300*(5620), 783–786.
61. Mokkapati, S., & Jagadish, C., (2009). III-V compound SC for optoelectronic devices. *Materials Today, 12*(4), 22–32.
62. Murad, A. R., Iraqi, A., Aziz, S. B., Abdullah, S. N., & Brza, M. A., (2020). Conducting polymers for optoelectronic devices and organic solar cells: A review. *Polymers, 12*(11), 2627.
63. Oh, S. J., Berry, N. E., Choi, J. H., Gaulding, E. A., Paik, T., Hong, S. H., & Kagan, C. R., (2013). Stoichiometric control of lead chalcogenide nanocrystal solids to enhance their electronic and optoelectronic device performance. *ACS Nano, 7*(3), 2413–2421.
64. Pauporté, T., & Lincot, D., (2000). Electrodeposition of semiconductors for optoelectronic devices: Results on zinc oxide. *Electrochimica Acta, 45*(20), 3345–3353.
65. Perebeinos, V., & Avouris, P., (2007). Exciton ionization, Franz–Keldysh, and stark effects in carbon nanotubes. *Nano Letters, 7*(3), 609–613.
66. Perebeinos, V., Tersoff, J., & Avouris, P., (2004). Scaling of excitons in carbon nanotubes. *Physical Review Letters, 92*(25), 257402.
67. Perzon, E., Wang, X., Admassie, S., Inganäs, O., & Andersson, M. R., (2006). An alternating low band-gap polyfluorene for optoelectronic devices. *Polymer, 47*(12), 4261–4268.
68. Pham, H. D., Xianqiang, L., Li, W., Manzhos, S., Kyaw, A. K. K., & Sonar, P., (2019). Organic interfacial materials for perovskite-based optoelectronic devices. *Energy & Environmental Science, 12*(4), 1177–1209.
69. Phillips, J. M., Burrows, P. E., Davis, R. F., Simmons, J. A., Malliaras, G. G., So, F., & Vittitow, M. P., (2006). Basic research needs for solid-

state lighting. *Report of the Basic Energy Sciences Workshop on Solid-State Lighting*. DOESC (USDOE Office of Science (SC)).

70. Piprek, J., (2016). Introduction to optoelectronic device principles. In: *Introduction to Organic Electronic and Optoelectronic Materials and Devices* (pp. 25–46). CRC Press.
71. Pop, E., Mann, D., Cao, J., Wang, Q., Goodson, K., & Dai, H., (2005). Negative differential conductance and hot phonons in suspended nanotube molecular wires. *Physical Review Letters, 95*(15), 155505.
72. Qiao, X., & Ma, D., (2020). Nonlinear optoelectronic processes in organic optoelectronic devices: Triplet-triplet annihilation and singlet fission. *Materials Science and Engineering: R: Reports, 139*, 100519.
73. Rakić, A. D., Djurišić, A. B., Elazar, J. M., & Majewski, M. L., (1998). Optical properties of metallic films for vertical-cavity optoelectronic devices. *Applied Optics, 37*(22), 5271–5283.
74. Ruini, A., Caldas, M. J., Bussi, G., & Molinari, E., (2002). Solid state effects on exciton states and optical properties of PPV. *Physical Review Letters, 88*(20), 206403.
75. Sahoo, D., & Naik, R., (2022). GSST phase change materials and its utilization in optoelectronic devices: A review. *Materials Research Bulletin, 148*, 111679.
76. Saito, R., Dresselhaus, G., & Dresselhaus, M. S., (2000). Trigonal warping effect of carbon nanotubes. *Physical Review B, 61*(4), 2981.
77. Shen, D., Chen, W. C., Lo, M. F., & Lee, C. S., (2021). Charge-transfer complexes and their applications in optoelectronic devices. *Materials Today Energy, 20*, 100644.
78. Sierros, K. A., Morris, N. J., Ramji, K., & Cairns, D. R., (2009). Stress–corrosion cracking of indium tin oxide coated polyethylene terephthalate for flexible optoelectronic devices. *Thin Solid Films, 517*(8), 2590–2595.
79. Simmons, J. M., In, I., Campbell, V. E., Mark, T. J., Léonard, F., Gopalan, P., & Eriksson, M. A., (2007). Optically modulated conduction in chromophore-functionalized single-wall carbon nanotubes. *Physical Review Letters, 98*(8), 086802.
80. Singh, T. B., Sariciftci, N. S., & Grote, J. G., (2009). Bio-organic optoelectronic devices using DNA. *Organic Electronics*, 73–112.

81. Spataru, C. D., Ismail-Beigi, S., Benedict, L. X., & Louie, S. G., (2004). Excitonic effects and optical spectra of single-walled carbon nanotubes. *Physical Review Letters, 92*(7), 077402.

82. Stewart, D. A., & Léonard, F., (2004). Photocurrents in nanotube junctions. *Physical Review Letters, 93*(10), 107401.

83. Stewart, D. A., & Léonard, F., (2005). Energy conversion efficiency in nanotube optoelectronics. *Nano Letters, 5*(2), 219–222.

84. Sun, X., Liu, Y., Xu, X., Yang, C., Yu, G., Chen, S., & Zhu, D., (2005). Novel electroactive and photoactive molecular materials based on conjugated donor– acceptor structures for optoelectronic device applications. *The Journal of Physical Chemistry B, 109*(21), 10786–10792.

85. Sweeney, S. J., & Mukherjee, J., (2017). Optoelectronic devices and materials. In: *Springer Handbook of Electronic and Photonic Materials* (pp. 1–1). Springer, Cham.

86. Tao, L., Chen, Z., Li, Z., Wang, J., Xu, X., & Xu, J. B., (2021). Enhancing light-matter interaction in 2D materials by optical micro/nano architectures for high-performance optoelectronic devices. *InfoMat, 3*(1), 36–60.

87. Tersoff, J., Freitag, M., Tsang, J. C., & Avouris, P., (2005). Device modeling of long-channel nanotube electro-optical emitter. *Applied Physics Letters, 86*(26), 263108.

88. Valentini, L., Mengoni, F., Armentano, I., Kenny, J. M., Ricco, L., Alongi, J., & Mariani, A., (2006). Enhancement of photoelectrical properties in polymer nanocomposites containing modified single-walled carbon nanotubes by conducting dendrimer. *Journal of Applied Physics, 99*(11), 114305.

89. Wang, F., Dukovic, G., Brus, L. E., & Heinz, T. F., (2005). The optical resonances in carbon nanotubes arise from excitons. *Science, 308*(5723), 838–841.

90. Wang, V., & Kuriyan, A. E., (2020). Optoelectronic devices for vision restoration. *Current Ophthalmology Reports, 8*(2), 69–77.

91. Wei, J., Zhu, H., Wu, D., & Wei, B., (2004). Carbon nanotube filaments in household light bulbs. *Applied Physics Letters, 84*(24), 4869–4871.

92. Wośko, M., Paszkiewicz, B., Piasecki, T., Szyszka, A., Paszkiewicz, R., & Tłaczała, M., (2005). Applications of functionally graded materials in optoelectronic devices. *Optica Applicata, 35*(3).

93. Wu, S., Chen, Z., Yip, H. L., & Jen, A. K. Y., (2021). The evolution and future of metal halide perovskite-based optoelectronic devices. *Matter, 4*(12), 3814–3834.
94. Xie, C., Wang, Y., Zhang, Z. X., Wang, D., & Luo, L. B., (2018). Graphene/semiconductor hybrid heterostructures for optoelectronic device applications. *Nano Today, 19*, 41–83.
95. Xu, H., Yin, L., Liu, C., Sheng, X., & Zhao, N., (2018). Recent advances in biointegrated optoelectronic devices. *Advanced Materials, 30*(33), 1800156.
96. Yang, R., Xu, Y., Dang, X. D., Nguyen, T. Q., Cao, Y., & Bazan, G. C., (2008). Conjugated oligoelectrolyte electron transport/injection layers for organic optoelectronic devices. *Journal of the American Chemical Society, 130*(11), 3282, 3283.
97. Yang, Z., & Hao, J., (2019). Recent progress in 2D layered III–VI semiconductors and their heterostructures for optoelectronic device applications. *Advanced Materials Technologies, 4*(8), 1900108.
98. Yin, D., Jiang, N. R., Liu, Y. F., Zhang, X. L., Li, A. W., Feng, J., & Sun, H. B., (2018). Mechanically robust stretchable organic optoelectronic devices built using a simple and universal stencil-pattern transferring technology. *Light: Science & Applications, 7*(1), 1–8.
99. Yu, J. C., Jang, J. I., Lee, B. R., Lee, G. W., Han, J. T., & Song, M. H., (2014). Highly efficient polymer-based optoelectronic devices using PEDOT: PSS and a GO composite layer as a hole transport layer. *ACS Applied Materials & Interfaces, 6*(3), 2067–2073.
100. Yun, J., (2017). Ultrathin metal films for transparent electrodes of flexible optoelectronic devices. *Advanced Functional Materials, 27*(18), 1606641.
101. Zeng, Q., & Liu, Z., (2018). Novel optoelectronic devices: Transition-metal-dichalcogenide-based 2D heterostructures. *Advanced Electronic Materials, 4*(2), 1700335.
102. Zhang, Y., Wu, J., Aagesen, M., & Liu, H., (2015). III–V nanowires and nanowire optoelectronic devices. *Journal of Physics D: Applied Physics, 48*(46), 463001.
103. Zhang, Z., Liu, Y., Liu, J., & Zhu, N., (2015). Packaging investigation of optoelectronic devices. *Journal of Semiconductors, 36*(10), 101001.

104. Zhao, H., & Mazumdar, S., (2004). Electron-electron interaction effects on the optical excitations of semiconducting single-walled carbon nanotubes. *Physical Review Letters, 93*(15), 157402.
105. Zhao, J., Chi, Z., Yang, Z., Chen, X., Arnold, M. S., Zhang, Y., & Aldred, M. P., (2018). Recent developments of truly stretchable thin film electronic and optoelectronic devices. *Nanoscale, 10*(13), 5764–5792.
106. Zhong, H., Lai, H., & Fang, Q., (2011). New conjugated triazine based molecular materials for application in optoelectronic devices: Design, synthesis, and properties. *The Journal of Physical Chemistry C, 115*(5), 2423–2427.
107. Zhou, K. L., Han, C. B., Li, C. F., Jiu, J., Yang, Y., Li, L., & Suganuma, K., (2018). Highly stable transparent conductive electrodes based on silver–platinum alloy-walled hollow nanowires for optoelectronic devices. *ACS Applied Materials & Interfaces, 10*(42), 36128–36135.
108. Zou, J., & Zhang, Q., (2012). Fundamental properties of carbon nanotubes. *Q. Zhang, Carbon Nanotubes and their Applications*, 1–30.

CHAPTER 5

ORGANIC SEMICONDUCTORS FOR OPTICAL APPLICATIONS

CONTENTS

5.1. INTRODUCTION

Organic semiconductors (OSC) were the subject of much research as potential new material for use in optoelectronics during the last two decades. They can integrate the straightforward manufacturing of plastics with both the functionality of electrical devices and the efficacy of light emission. They have found use in a wide variety of optoelectronics, such as organic light-emitting diodes (OLEDs) photo-voltaic, transistors, optical communications (Tang and VanSlyke, 1987; Amarasinghe et al., 2009), and lasers. Their one-of-a-kind mix of features has made it possible to develop novel devices geometries and operating methodologies, which has led to the development of things like printed optoelectronic devices. The use of OLEDs, in light-emitting screens has been the primary focus of research. OLEDs are now commercially available, and the significant success that they have achieved has made it a promising ground of study to investigate the wider application of these components in the creation of other light-emitting machines (Spanggaard and Krebs, 2004).

OSC have a lot of promise for use in the development of new kinds of laser gain mediums. These have power sources with four levels, and the excitation and fluorescence bands are separated from one another by vibrational and architectural relaxations. High optical transitions (peak absorption of 105 cm^{-1} and emissions cross-section 10^{-16} cm^2) offer the potential for a significant optical gain in extremely tiny devices. This potential is referred to as the optically gain density. Superconductors are bandwidth emitters, much like traditional laser dyes. However, the emission of semiconductor materials may be adjusted throughout the visible light spectrum by making very few adjustments to their chemical composition. On the other hand, in contrast to traditional dyes, they exhibit very minimal intensity quenching and may have fluorescence emission efficiency in unadulterated films that are more than 90%. As a consequence of this, OSC are just an excellent choice for the development of tiny tunable lasers as well as broadband optical communications (Burroughes et al., 1988; Lawrence et al., 2002).

At the moment, every organic semiconductor laser and optically amplifier is visually pumped, which means that additional light sources are required to stimulate the OSC. Pulsed microchip lasers now are frequently employed to drive widely adjustable organic distribution feedback lasers. There have been significant advancements made in the development of small systems, and these advancements have been positive. In addition to this, there have been other examples of organic laser diodes pumped by nitride laser diodes, and even an LED-pumped polymers laser (Heliotis et al., 2004). In addition, these materials have semiconducting characteristics, which opens the door to the possibility of semiconducting lasers with amplifiers that are electronically pumped. Since organic lasers were the subject of a wealth of research and analysis over the years, this chapter will concentrate on the more recent topic of organic semiconductors optical amplifiers (OSOAs) (Moses, 1992; Zhang, 2000).

Fiber optics are equipment that is used in communication systems because of their usefulness in amplifying the transmitted signal in such a single pass through the device. They are necessary because the dividing of the signal as well as the transmission lines both reduce its intensity, and these losses must be compensated for. At the moment, silica fibers are used as the foundation for broad-range communication systems, and these networks are capable of transmitting at high available bandwidth of much more than one gigabit per second. These very high bandpass filters are not yet accessible to private residences or commercial establishments (Dıaz-Garcıa et al., 1997). because coaxial cables (perverted pair or fiber cable), that are used for transmissions from far stations to local area networks (LANs), have such narrow bandwidth, therefore, producing a bottleneck at remote sites. The installation of silica fiber networks between distant units and LANs would not have been cost-effective since it would need the use of thousands of fibers, which would drive up the cost to an extremely high level. Polymer fiber optics, also known as POFs, are an affordable option that boasts a wide throughput while maintaining a low cost. They are easy to install and may be done so on the premises themselves. A bundle of polyolefin fibers (POF) is much more flexible than just a network of silica fibers, and so as a result, it is ideal for the severe environments that are seen in metropolitan areas. Because the wavelength of OSOAs' emissions is so closely suited to that of POF's transmitting windows, the two technologies might be interoperable with one another (McGehee and Heeger, 2000).

The majority of colors and linked polymers have optical changes that are located within the visible part of the spectrum. As a result, the vast majority of organic light-emitting technologies also function within the visible spectrum. The Peierls distortion restricts the lowest bandgap energy that could be attained; a conjugated polymer has the potential to become a one-dimensional (1D) material, but this type of metal is unpredictable to a dimerization, which creates an energy deficit and therefore results in semiconducting actions. This excitation wavelength may be increased (but only to a limited extent) by combining with substances including erbium tris(quinoline) or semiconductors quantum dots. However, the effectiveness of this process is restricted. Polymer nanocomposites have many advantages over dye-doped polymeric materials as solid-state amplifiers, along with much greater chromophore concentration levels that could be accomplished without intensity quenching, which allows for high photoluminescence (PL) produced even now in neat films and has the potential for future electrical stimulation. Dye-doped polymeric materials have become the subject of a significant amount of research as strong amplifiers (Scherf et al., 2001; Vasdekis et al., 2005).

5.2. ELECTRONIC STRUCTURE AND OPTICAL PROPERTIES

The semi-conductive characteristics of semiconductor materials, like those of their inorganic materials, resulting from the overlapping of atomic orbitals (AOs). Three out of the four electron pairs in carbon normally form covalent–bonds with nearby atoms inside the planes of coiled molecules, which alternate single and double bonds. The fourth electron sits inside an orbital called a pz that is parallel to the particle's plane. As seen in Figure 5.1, the overlapping of such orbitals results in covalently r-bonding and electrons delocalization throughout the molecules, which in turn results in semiconducting electrical characteristics. Pz AOs may create bonded (r) or anti-bonding (r*) molecular orbital angular momentum. Two electrons typically occupy the bonding r-orbital within the occupied molecular orbital (homo orbital (HOMO). Some of which may be excited to an anti-bonding r* orbital known as the lowest unoccupied molecular (LUMO). The organic substances' lowest frequency absorption and luminescence capabilities are determined by transitions among HOMO and LUMO. Transportation holes occurs by hopping between HOMOs of neighboring mole- cules, whereas transport of electrons is by hopping be- tween the LUMOs of neighboring molecules. (Vasdekis et al., 2006).

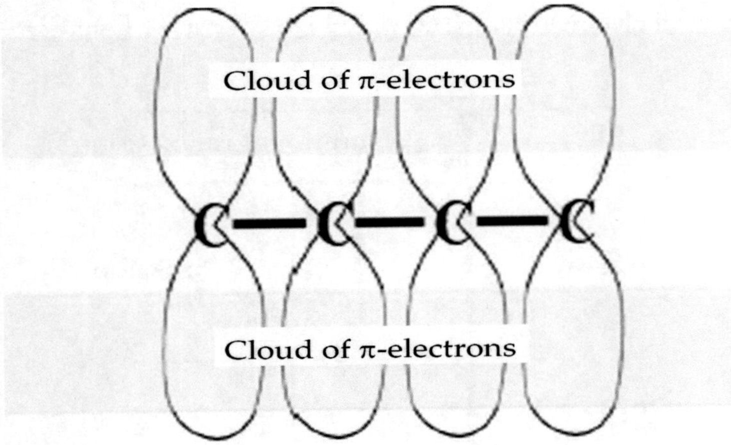

Figure 5.1. Figure depicts the pz-orbitals above and below the polyacetylene plane. Molecular r-orbitals are produced by the overlap of pz-orbitals and are conjugated throughout the polymer chain.

Source: https://www.masterorganicchemistry.com/2017/02/28/pi-molecular-orbitals-of-butadiene/.

Transportation of electrons, on the other hand, happens via hopping between both the LUMOs of nearby molecules and not between HOMOs of adjacent molecules. Whereas the HOMO and LUMO are comparable to the valence band and conduction band of semiconductor materials, band-like trans-migration is only conceivable in the most organized semiconductor materials. Hoping transport is caused by the fact that most superconductors are substantially more disorganized than inorganic semiconductors.

Organic laser pigments are linked compounds that, when excited optically in solutions or when spread in such an inert matrix, may produce optical gain. Polymer nanocomposites offer the advantages of being able to produce waveguides with ease, having good PL emission outputs just like neat films, and having the potential for electrically pumped in future (Xia et al., 2005; Rabe et al., 2006).

The material may capture a photon with energy equivalent to or higher than just the energy difference between both the HOMO and LUMO, which will stimulate a molecule from the quantum mechanical S_0 to a protonated excited singlet state. S_1 stands for the shortest singlet excited singlet state, whereas S_2, S_3, etc., are higher rising singlets. There are several vibrational levels for each electrical state (shown in Figure 5.2 as 0, 1, and 2).

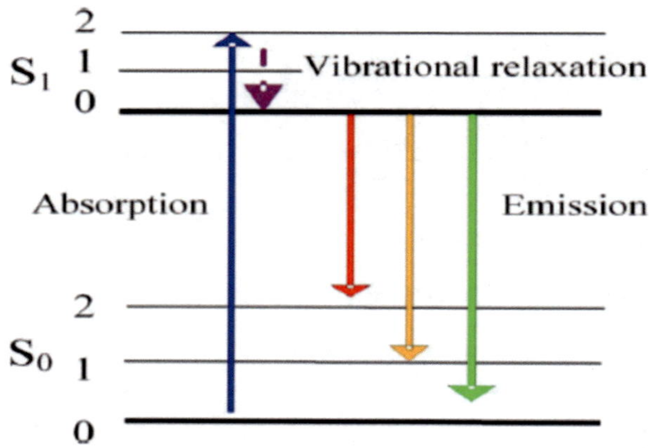

Figure 5.2. Photon emission and absorption are caused by optical transitions between the ground (S_0) state and the lowest singlet excited state (S_1) (solid arrows). An arrow with a dash indicates the radiation-free process of vibrational relaxation in S_1.

Source: *https://chem.libretexts.org/Bookshelves/Physical_and_Theoretical_Chemistry_Textbook_Maps/Supplemental_Modules_%28Physical_and_Theoretical_Chemistry%29/Spectroscopy/Electronic_Spectroscopy/Jablonski_diagram.*

Since the distance among excited states is substantially greater than the heat energy, practically all compounds are now in the zero-vibrating frequency of the S_0 state before absorbing light. Based just on photon energy, absorbing happens to one of the vibrational levels of a singlet excited singlet state. Following ingestion, relaxing to the S_1 state's zero vibration energy level typically takes place in less than one picosecond. From the lowest vibration frequency of the S_1 phase with each of the excited states of the quantum mechanical S_0, photon emission takes place (either spontaneously or in response to an external photon) (Zavelani et al., 2006).

Figure 5.3 shows to show high gain optic amplifier under solution (Figure 5.3(a)–(c)) and that in the crystalline structure (Figure 5.3(b), (d), and (e)), various conjugated biomolecules (four polymeric and a first-generation dendrimer) were utilized. The linked polymer OC1C10—PPV's passion for creating and absorption spectra are shown in Figure 5.4. Peaks at 560 and

600 nm, as well as shoulders at 650 nm, are seen in the emission spectra. These characteristics signify optically transitions first from the S_1 state's lowest vibration level to its excited states 0, 1, and 2, correspondingly. This pattern is obscured by non-homogeneous flattening in optical absorption (Riechel et al., 2001).

OSOAs, like some other substances, may be amplified by emitted photons.

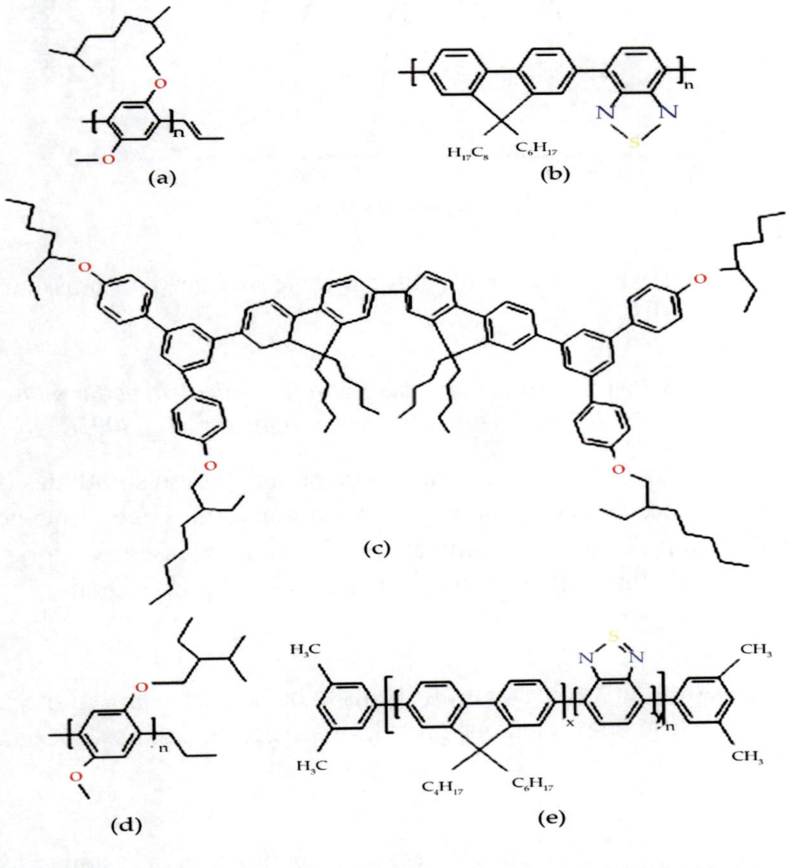

Figure 5.3. Several organic semiconducting substances' chemical compositions are employed in amplifiers: (a) OC1C10-PPV; (b) F8BT; (c) dendrimer of the first generation with a bisfluorene core; (d) MEH-PPV; and (e) ADS233YE.

Source: https://www.academia.edu/58466163/Semiconducting_Organic_Molecular_Materials.

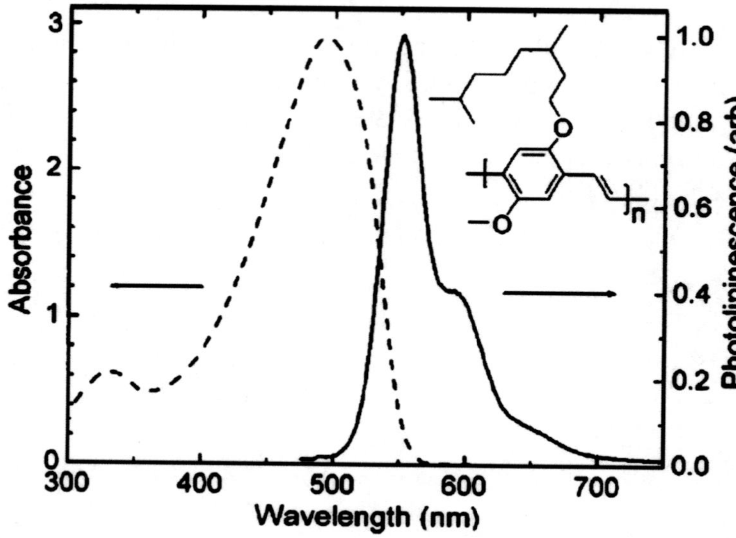

Figure 5.4. (Left) Absorption and (right) molecule structure and emission spectra of OC1C10-PPV.

Source: https://www.researchgate.net/figure/Chemical-structures-of-a-OC-1-C-10-PPV-b-capped-TOPO-TOP-nanocrystal-and-c_fig1_230958746.

The shift from S_1 to S_0 is crucial. An incoming photon stimulates it, and as a consequence, an additional photon with much the same frequency, as well as direction even as the stimulating photon, is released. With range z, the optic signal's intensity increases exponentially for tiny signals.

$$I = I_0 \exp^{(g-\alpha)z}$$

wherein; g is the gains coefficient; I_0 seems to be the starting brightness; and 2 is the absorption coefficient as a result of scattering and absorption. This latter is shown by:

$$g = \sigma N$$

where; N represents the S_1 state's density of population and stands for the gains cross-section. Transient absorption spectroscopic or amplification spontaneous emission (ASE) observations utilizing the varying stripes length (VSL) approach, as shown by McGehee et al. may be used to determine these parameters. The existence of ASE suggests that g is bigger than, and also that the gain exceeds the absorbing. For example, Xia et al. applied the VSL technique to many conductive polymers.

Red-F, poly(9,9-dioctylfluorene), and poly(9,9-dioctylfluorene-co-benzothiadiazole) (F8BT) (Turnbull et al., 2003; Karnutsch et al., 2007) are examples of these compounds. The measurements showed waveguiding losses of 3.5, 3.2, and 7.6 cm^1, gain cross-sections of 10^{-16} cm^2, net gains of 74, 24, and 22 cm^1, and gain cross-sections of 10^{-16} cm^2. These numbers are in the same range as those recorded for those other conductive polymers. Gain is governed by the optical communication systems discussed in this study (Sakata and Takeuchi, 2008).

$$\pm(\quad) = 10\log\left[\frac{I_{\pm} - I}{I_{\pm}}\right]$$

wherever; I_{on} remains the strength of the enhanced indication; I_b is related to luminescence (characteristically I_b G 0.01 of I_{on}), I_{off} remains the sign strength without a pump.

5.3. SOLUTION-BASED AMPLIFIERS

5.3.1. Conjugated Polymers as Gain Media

Transient absorbance and VSL methods have been used to detect the existence of gains in optical transmission solutions containing semiconductor materials, but it wasn't until later that a linked polymer was found to exhibit substantial amplified. Laser pulses in the ultraviolet (UV) were used to excite solutions of poly[2-methoxy-5-(3,' 7'-dimethyl octyl oxy)-paraphenylenevinylene] (OC1C10-PPV; Figure 5.4) in chlorobenzene in a 1 cm quartz cuvette. The cylinder lens focused the pumping beam into such a stripe. A voltage meter was positioned after samples were used to track the poor signal pulse's development; a few of the findings are displayed in Figure 5.5. A tuning frequency of around 50 THz was used to generate a maximum benefit of 30–43 dB. This work demonstrated that polymer nanocomposites are responsible to increase light pulses across a wide range of wavelengths and by many magnitudes. In actuality, the dye beam's tunable range, which was utilized to produce the measured voltage that's been amplified, rather than the polymer membrane, was what restricted the claimed different wavelengths. According to Figure 5.5, density has an impact on gain as well. At small concentrations, not enough pump light energy is absorbed in the transmission beam's region, and while at high levels, the pump radiation is absorbed in a region much smaller than just the signal beam. As a result, the transitional concentration has the maximum gain (Yang et al., 2008).

Figure 5.5 displays the observed gain fluctuation with energy input Ein (b). The solid line theoretically fits the optic amplifier's gain (Yang et al., 2008).

$$G = \frac{A}{C_3} In\left[1 + G_0[\exp\left(\frac{C_3}{A}\right) - 1]\right]$$

where;

$$C_3 = \frac{E_{in}(\sigma_{abs} + \sigma_{se})}{hv_{in}}$$

and where; A is the cross-sectional surface of a signal; σ_{abs}, σ_{se} is the absorbance and stimulation emissions cross-section; h is Planck's constant, v_{in} seems to be the velocity of the signal, and G_0 is the tiny signal gain or the unsaturated obtain achieved with poor signal energy. The energy input during which gain equals 25% of a G_0 value is known as the saturated energy, or E_{sat}. E_{sat} = 1 J and a minor signal gain of 44 dB are found from the fit.

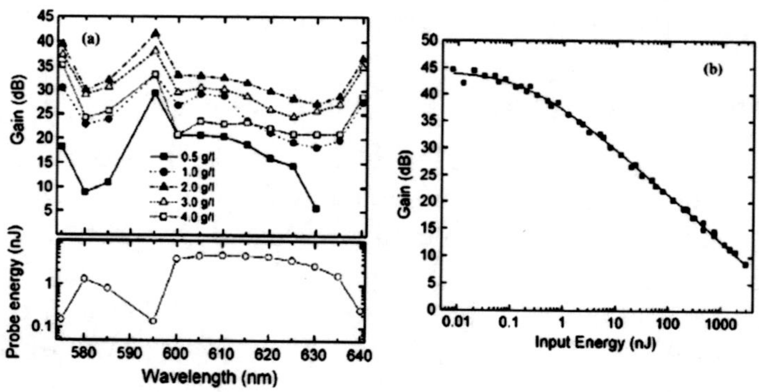

Figure 5.5. (a) Gain for five distinct concentrations of OC1C10-PPV in chlorobenzene solution as a function of signal wavelength. The signal pulse energy is shown on the bottom panel. (b) Gain about signal energy at a 600 nm signal wavelength.

Source: https://www.degruyter.com/document/doi/10.1515/9783110304558-006/pdf.

The gain cross-section was calculated to be = (5.3 0.6)× 10^{-17} cm^2 using both (2) (calculating only in the range of low pump energy densities) and (5) (assuming that the absorption at the probing wavelength to be negligible, hence $\sigma_{abs} = 0$), which is expected for high-gain conjugated polymers

Owing to a signal pulse's reduced energy due to the location at the border of the dyed laser generation's frequency band, the gain increases at the edges of a frequency band (575, 640 nm). The linked copolymer F8BT, seen in Figure 5.3(b), has already been employed as a host material in polymeric lasers and also solutions and strong amplifiers (Monroy et al., 2003).

Over the spectral region from 535 to 570 nm, solution amplifiers displayed gains over 30 dB (Figure 5.6(a)). Low ground state absorbance in the stimulated sample may explain an association with elevated gain towards wavelengths shorter with large levels. The gain fluctuation with the signals pulse frequency for a level of 3 mg/ml is shown in Figure 5.6(b). Above Eqn. provided a modest need to recognize 28 dB and saturated signal energy of 0.6 J when it was fitted to experimental results. The gain cross-section, which has been calculated first from fit and is similar to previous values published was found to be (1.7 0.1) 10^{-16} cm^2 (Peierls and Peierls, 1955).

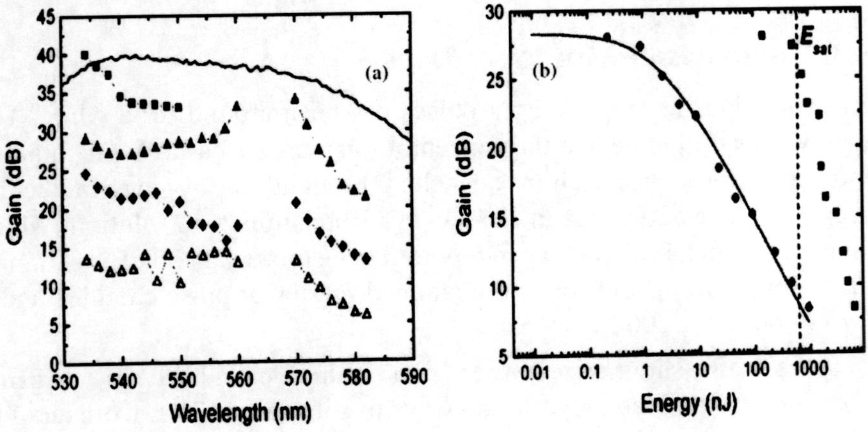

Figure 5.6. Gain characteristics of F8BT in the solution include: (a) gain as a function of signal wavelength at concentrations of 1.5 (open triangles), 3, 5, and 3.5 mg/ml. The steady-state photoluminescence spectrum is shown by the solid line. 350 J/pulse was the pump energy. (b) Gain as a function of the signal energy of the input (circles) and output (squares). The continuous line fits to (5). **The saturation signal energy Esat is indicated by the dotted line.**

Source: https://www.researchgate.net/figure/Amplifier-gain-as-a-function-of-wavelength-for-a-range-of-polymer-concentrations-15_fig2_234925491.

5.3.2. Dendrimers

Additionally, linked dendrimers, a different form of organic semiconductor, have been shown to exhibit optical amplification. These branching macromolecules have a core, dendrons, and are adsorbed on the surface, and they may have a wide range of characteristics because of their modularity molecular design. A first-generation bisfluorene-cored dendrimer that had earlier shown promise for bright blue OLEDs was used to study the potential for optical amplifiers. This dendrimer, which is seen in Figure 5.3, is made up of 2 G1 triphenyl-based dendrons with 2-ethyl-hexyloxy surface groups attached to something like a bisfluorene unit (c). The dendrimer was shown to produce substantial optical amplification within deep blue as well as nearly UV parts of the spectrum when it was dissolved in toluene at a ratio of 0.31 g/l. The highest gains of 36 and 26 dB, respectfully, were attained with a signal wavelength of 420 and 390 nm, as well as the outcomes, are displayed in Figure 5.7. It was determined that the gained cross-section was equal to 3.4×10^{-18} cm^2 (Gillin and Curry, 1999).

5.3.3. Comparison with Laser Dyes

Using laser dyes to amplify light pulses has been around for a while. The key drawback would be that the concentration increase has to be maintained low to prevent concentration quenched brought on by dye molecular aggregation. The optical magnification in a Rhodamine 640 solutions with a 1 cm path length has already been reported to be more than 30 dB by Ramon et al. Gains of 30 dB or more were attained for signal pulse energies under 100 nJ (Coe et al., 2002).

Figure 5.8 summarizes the gain value calculated in organic semiconductor- and laser dye-based ways to solve amplifiers. Compared to laser dyes, superconductors have strong gain across a broader bandwidth. Organic semiconductors show high gain over a wider bandwidth than laser dyes.

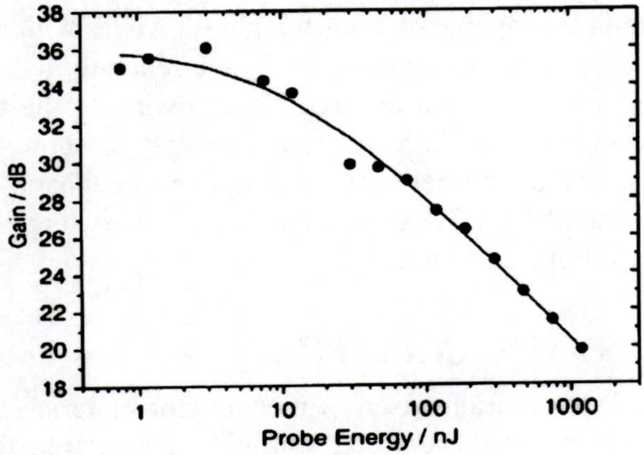

Figure 5.7. Gain dependency for a bisfluorene-cored conjugated dendrimer in solution at the 430 nm signal pulse energy.

Source: https://www.researchgate.net/figure/Amplifier-gain-as-a-function-of-wavelength-for-a-range-of-polymer-concentrations-15_fig2_234925491.

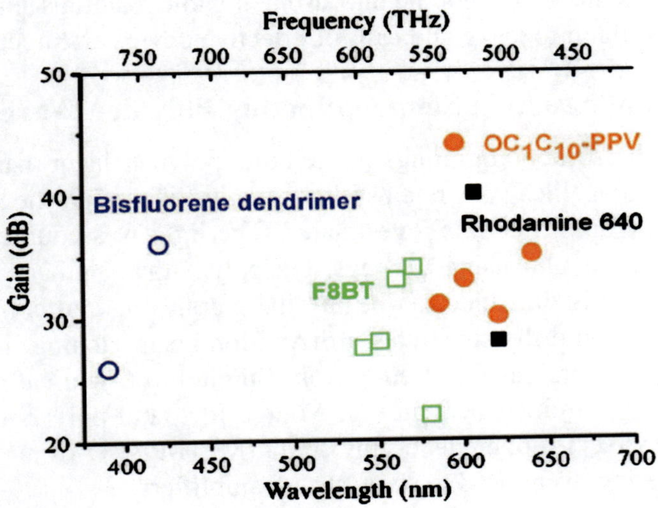

Figure 5.8. Maximum gain in solution amplifiers using a bisfluorene-cored first-generation dendrimer (open circles), F8BT (open squares), and OC1C10-PPV (open lines) as a function of signal wavelength (solid circles).

Source: https://pubs.acs.org/doi/10.1021/acs.jpcc.2c00648.

It shows the threshold energy for such signal pulse, shows that both kinds of materials are capable of amplifying laser waves with input energies up to 1 J. Organic semiconductors (OSC) can generate high voltage gain in such a smaller route length that laser dyes owing to the unavailability of concentrate quenching. Although the findings do show that polymer nanocomposites and dendrimers have promise for broadband amplification devices, every one of the OSOAs presented in this study is pulsed (Karplus and Porter, 1970; Pope and Swenberg, 1999).

5.4. SOLID-STATE AMPLIFIERS

The findings of the solution-based semiconductor materials amplifier are highly promising; however, a strong amplifier is preferred for real-world applications. This creates new difficulties:

- The pump, as well as probe lasers, overlap across a lower density than in the way to solve studies;
- Films' increased chromophore concentration might cause some intensity quenching;
- It is difficult to couple into strong organic materials since they are difficult to leave and call for effective device architectures.

5.4.1. Amplification in Semiconducting Polymer Waveguides

Utilizing slab diffraction gratings formed of a polymer layer that was spin-coated on over a silica substrate as illustrated in Figure 5.9, the viability of the polymeric amplifier was investigated. The light was coupled into and out of protective film using gratings. On polymethyl methacrylate, these were created utilizing electron-beam lithography, and afterwards, they were deposited into the substrates utilizing ion beam etching. Industrially accessible polymers such as Red-F from Sumitomo Chemical Co., F8BT and ADS233YE from American Dye Source Inc., and poly[2-methoxy-5-(2'-ethylhexyloxy)-p-phenylene-vinylene] (MEH-PPV) from American Dye Sources Inc. were used to show strong amplifiers.

In Figure 5.10, the outcomes of the MEH-PPV solid-state amplifiers, that demonstrated a tiny need to recognize up to 21 dB in a wavelength 1 mm long (Nie et al., 1998). While in low pump power density area, a gained cross-section of 4×10^{-17} cm^2 was attained utilizing, while gain saturates were seen at high pump concentrations. Gain saturation is caused by ASE with exciton-exciton annihilation, according to research on gain kinetics.

Exciton-exciton annihilation is just a process that results in the loss of one exciton when two excited electrons inside the annihilation radius combine to generate a higher excited electron. The constant was determined to be 3×10^{-9} cm^3/s, which is comparable to that seen in films of many other biopolymers. This is a bimolecular activity. Gain dependency on signals pulsed energy is seen in the inset of Figure 5.10 and shows that now the community inversion at modest pump concentrations supports amplification of changes in visual pulses up to 30 nJ in intensity. Additionally, this polymeric amplifier demonstrated a gain of 15 dB at 615 to 650 nm wavelengths, providing a gain spectrum of> 26 THz (Hayes et al., 1995; Eastham, 2000).

Figure 5.11 shows gain characteristics in Red-F copolymer slab waveguides. Only 0.3 mm in length and showing a gain of 18 dB, the largest waveguide employed in the research detailed in this study was short. This shows that now the diluted character of reduced chromophores in copolymers might enhance gain qualities since it is only lower overall than that seen in the 1 mm MEH-PPV waveguide. A recent investigation of gains in waveguides of the copolymer ADS233YE has validated this, demonstrating a maximum benefit of 32 dB in such a 1-mm-long waveguide (Yan et al., 1994). The promise of conjugated polymers for small optical communication systems is shown by these extremely high gains over extremely short distances (Yan et al., 1994).

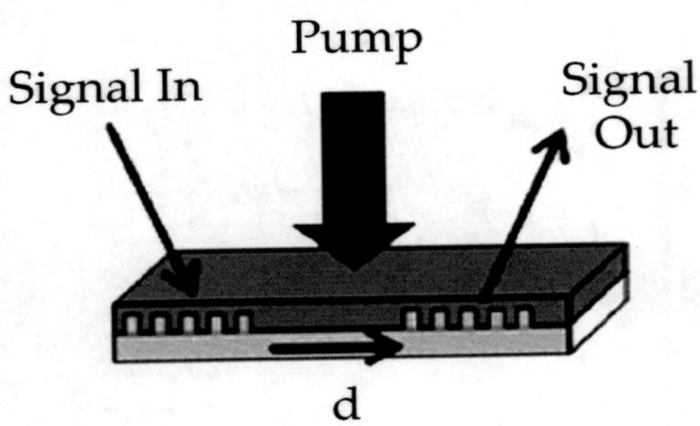

Figure 5.9. Slab waveguide polymer amplifier schematic with grating couplers.

Source: https://opg.optica.org/ao/abstract.cfm?uri=ao-36-36-9383.

5.4.2. Amplification of Multiple Pulses

It is crucial to look at the viability of amplification presidential since data transfer is accomplished utilizing light pulse streaming. This is difficult so because the gain lifespan of earlier conjugated polymer amplifiers was comparatively brief at high pump power density. Three pulses in such a 140 ps window in such a semi-conductive polymer waveguide have been used to investigate it.

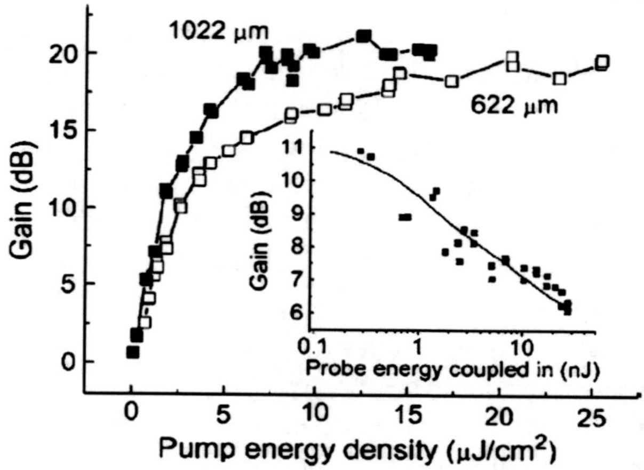

Figure 5.10. Gain dependency on pump power density in 1 to 0.6 mm MEH-PPV diffraction gratings. The input pulsed energy of the signal, which was at 630 nm, was consistent at 2 nJ. The main route is a theoretical fit utilizing as well as the inset depicts gain dependency on input optical pulse energy.

Source: https://opg.optica.org/oe/fulltext.cfm?uri=oe-23-25-31926.

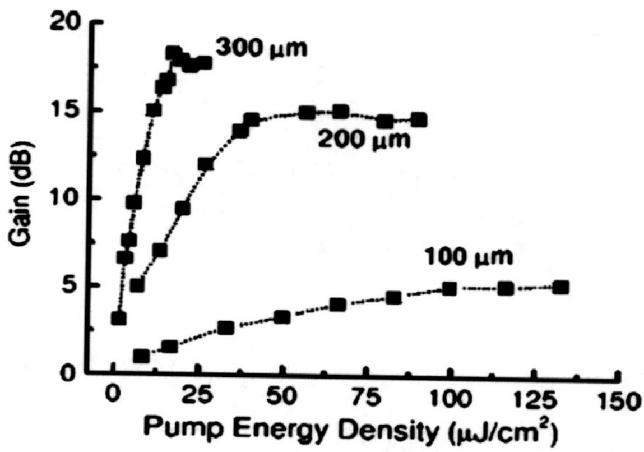

Figure 5.11. Gain variation in waveguides made of RedF copolymer that was 0.3, 0.2, and 0.1 mm long concerning pump energy density. The wavelength of the signal was 660 nm. The wavelength of the signal was 660 nm.

Source: https://www.mdpi.com/2227-9040/10/5/150/htm.

Results are shown in Figure 5.12(a), where the initial pulse exhibits a gain of 24 dB as well as the 2nd and 3rd pulses exhibit amplification of 18 and 13 dB, accordingly, as a consequence of the excited singlet state population's natural decay. Gain dependency on signals pulse power is shown in Figure 5.12(b), demonstrating that sometimes the amplification of triple pulses with input energies of up to 0.1 nJ is feasible. Although it hasn't been shown yet, utilizing a lengthy pump pulse might be able to amplify light pulses across a wider period (Schwartz et al., 1997).

5.4.3. Comparison with Dye and Rare-Earth Ion Doped Polymer Waveguides

Other than semiconductor materials, dye-doped, as well as rare-earth ion doped polymers waveguides and fibers, are all the other potential gain media that fit the transparent windows of the polymeric optical fiber (500–560 and 660 nm). The chromophores are embedded in an optical passive insulator matrix material in such systems.

Next, we look at the development of a dye-doped waveguide amplifier. As was already noted, organic waveguides are challenging to cleave, making end couplings into polymers challenging. By employing a ribbed waveguide construction, Reilly et al. were able to get around this. The 1.2 cm long and 120 m broad waveguide structure is constructed of a layer of germanium-doped silica that was placed on a silica substrate. Waveguides 10 m in depth were carved onto the doped silica layer. To create a 1-m-thick layer, the lasers dye Rhodamine 640 being spin-coated upon poly(methacrylate) (PMMA) at a concentration of 1% by weight. The signal's length was 625 nm, and it has been pumped using 10 ns pulsed at 575 nm with a sampling frequency of 10 Hz. The photons were linked into a silica waveguide and then merged. The aqueous compound with a higher index of refraction received light from the waveguide. A signal-to-noise proportion (SNR) of 9–16 dB was used to produce a maximum benefit of 26 dB with such signal energy of 4 nJ. Reilly et al. employed the photoresist SU8 spin layered over a 7 m-thick silica shell and structured into 100 m extended functionality as a secondary dye-doped polymeric waveguide system. Low optical waveguides might be produced simply using SU8. An evanescent wave was linked into the protective film and amplified while the light was constrained within SU8 waveguides. An SNR in the region of 15 to 4 dB was required to achieve the maximum benefit of 16 dB. Prometheus 650 (P650) embedded in a PMMA matrix was used by Lam et al. to create a pigment-doped solid phase amplifier (Xia et al., 2003). There was a film of an active ingredient.

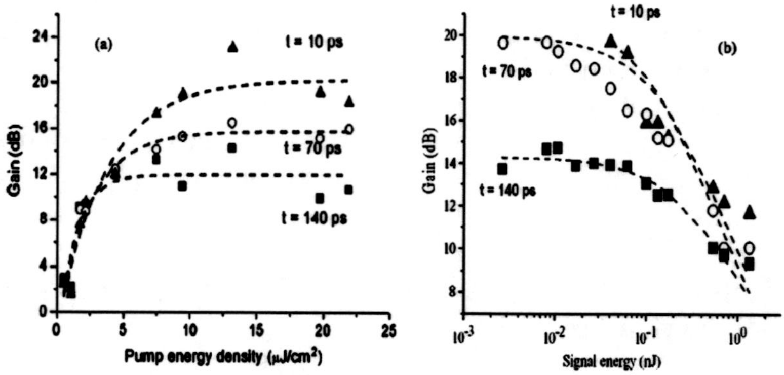

Figure 5.12. Multiple pulses are amplified in an F8BT amplifier. (a) Gain dependency on pump energy density in a 1 mm long waveguide with signal energy of 0.04 nJ. The wavelengths of the signal and pump were 497 nm and 580 nm, respectively. (b) Gain sensitivity to signal energy (symbols) in a 1 mm waveguide.

Source: https://www.researchgate.net/publication/235709061_Raman_amplification_of_optical_pulses_in_silicon_nanowaveguides_Impact_of_spectral_broadening_of_pump_pulses.

The highest gains ranging up to 25 dB were attained for just a resonant wavelength of 616 nm when spin-coated on such a substrate.

The gain ratios have been established using VSL observations, as was previously noted. Many measurements involving dyes have already been made, including the following. 2,7-(1,3-dihydro-1,3,3-trimethyl-2H-indol-2-ylidene)-1,3,5-heptatrienyl]-1,4-bis[2-[4-[N,N-di(p-tolyl)amino]phenyl]vinyl]benzene (B2080), Rhodamine 640, Coumarin 540, and 1,3,3-trimethyl-3H-indolium iodide (NK-125) are other examples. Gain ratios of more than 50 cm^{-1} have been recorded.

Both dye-doped fibers amplifiers and fiber lasers have been the subject of research through. We shall concentrate on dye-doped optic fiber amplifiers, although this is a study of amplifiers. Such devices were created by doping POF or grading index POF (Schwartz et al., 1997) using dye materials. Since optically gain happens on atomic transition, the resulting linewidth is substantially shorter than that of molecule materials, resulting in a reduced gain bandwidth. To make rare-earth ions more compatible with such a polymer matrix and therefore enable doping at greater concentrations than that in silica to create compact circuits, rare-earth ions may be enclosed with organic linkers. A step-index polymeric fiber optic with such a 30-cm length that's been doped using europium showed a gain of 5.7 dB at 613 nm with the use of fiber amplifiers made of erbium-doped polymers, increases of up to 7 dB/cm have already been attained. Neodymium, one more rare earth material, has already been studied and provided an increase of 1.6 dB/cm. A description of the gain in organic components amplifiers is shown in Figure 5.12. When compared to doped fiber amplifiers, superconductors have a substantially better gain each unit length and are therefore smaller. Since the goal of this work is to have a summary of gain qualities, transmissions, and couplings losses will not be included in that comparison (Heliotis et al., 2003).

5.5. CONCLUSION

For a little more than 10 years, especially in the following and ASE has already been studied in conductive polymers conjugated polymer films. Semiconducting linked polymer amplification devices, on the other hand, are a considerably more recent development. In diluted as well as concentrated films, polymers have shown gains by the unit length of 21–60 dB/mm. Such values include up to 80 times as high as just those produced using rare earth minerals amplifiers as well as up to 30 times greater than the maximum gain per length l reported in a dye-doped polymeric amplifier. This demonstrates that substantially smaller amplifiers may be created because of the small concentrations quenched in conjugated polymers. This may result in integrated optical amplifiers that are small and very suitable with POF.

Moles through. Many bioactive molecules have indeed been employed, including: 4-Dimethyl-2-cyclohexyl-6-(p-dimethylaminostyryl) – oxazine 4 perchlorate (O4PC), 4H-pyran (DCM).

Rhodamine has been the most widely used dye, followed by Rhodamine 6G with rhodamine B (RB). Pump-probe tests with a signal wavelength closer to the low optical window of PMMA were used to show amplification. The highest gains within the 20–35 dB/m range have indeed been attained. Rajesh et al. used Rhodamine 6G at such a dye concentration of 1×10^{-3} M to achieve a very significant amplified of 18 dB in such an 8-cm-long fibers amplifier. The numbers represent the gain potential in cm of fiber length of dye-doped materials.

The buildup of triplet seismic excitation, which produces losses, involves continuous waveform or high repetition rates operation troublesome for lasers dyes. Rare-earth particles have indeed been employed as a substitute as a gain medium in polymer fiber optics. They expand upon the accomplishments of erbium-doped silica-based fiber amplifiers, that have been productively used during silica broadband networks.

REFERENCES

1. Amarasinghe, D., Ruseckas, A., Turnbull, G. A., & Samuel, I. D., (2009). Organic semiconductor optical amplifiers. *Proceedings of the IEEE, 97*(9), 1637–1650.
2. Amarasinghe, D., Ruseckas, A., Vasdekis, A. E., Goossens, M., Turnbull, G. A., & Samuel, I. D. W., (2006). Broadband solid state optical amplifier based on a semiconducting polymer. *Applied Physics Letters, 89*(20), 201119.
3. Amarasinghe, D., Ruseckas, A., Vasdekis, A. E., Turnbull, G. A., & Samuel, I. D. W., (2008). Picosecond gain switching of an organic semiconductor optical amplifier. *Applied Physics Letters, 92*(8), 63.
4. Amarasinghe, D., Ruseckas, A., Vasdekis, A. E., Turnbull, G. A., & Samuel, I. D. W., (2007). Amplification of optical pulse sequences at a high repetition rate in a polymer slab waveguide. *Applied Physics Letters, 91*(1), 011105.
5. Burroughes, J. H., Jones, C. A., & Friend, R. H., (1988). New semiconductor device physics in polymer diodes and transistors. *Nature, 335*(6186), 137–141.
6. Coe, S., Woo, W. K., Bawendi, M., & Bulović, V., (2002). Electroluminescence from single monolayers of nanocrystals in molecular organic devices. *Nature, 420*(6917), 800–803.
7. Denton, G. J., Tessler, N., Harrison, N. T., & Friend, R. H., (1997). Factors influencing stimulated emission from poly (p-phenylenevinylene). *Physical Review Letters, 78*(4), 733.
8. Dıaz-Garcıa, M. A., Hide, F., Schwartz, B. J., McGehee, M. D., Andersson, M. R., & Heeger, A. J., (1997). "Plastic" lasers: Comparison of gain narrowing with a soluble semiconducting polymer in waveguides and microcavities. *Applied Physics Letters, 70*(24), 3191–3193.
9. Eastham, J., (2000). Designing out failure in composites: One day seminar. *Materials Technology, 15*(1), 96–98.
10. Gillin, W. P., & Curry, R. J., (1999). Erbium (III) tris (8-hydroxyquinoline) (ErQ): A potential material for silicon compatible 1.5 μm emitters. *Applied Physics Letters, 74*(6), 798–799.

11. Hayes, G. R., Samuel, I. D. W., & Phillips, R. T., (1995). Exciton dynamics in electroluminescent polymers studied by femtosecond time-resolved photoluminescence spectroscopy. *Physical Review B, 52*(16), R11569.
12. Heliotis, G., Bradley, D. D. C., Goossens, M., Richardson, S., Turnbull, G. A., & Samuel, I. D. W., (2004). Operating characteristics of a traveling-wave semiconducting polymer optical amplifier. *Applied Physics Letters, 85*(25), 6122–6124.
13. Heliotis, G., Xia, R., Bradley, D. D. C., Turnbull, G. A., Samuel, I. D. W., Andrew, P., & Barnes, W. L., (2004). Two-dimensional distributed feedback lasers using a broadband, red polyfluorene gain medium. *Journal of Applied Physics, 96*(12), 6959–6965.
14. Heliotis, G., Xia, R., Bradley, D. D. C., Turnbull, G. A., Samuel, I. D. W., Andrew, P., & Barnes, W. L., (2003). Blue, surface-emitting, distributed feedback polyfluorene lasers. *Applied Physics Letters, 83*(11), 2118–2120.
15. Karnutsch, C., Stroisch, M., Punke, M., Lemmer, U., Wang, J., & Weimann, T., (2007). Laser diode-pumped organic semiconductor lasers utilizing two-dimensional photonic crystal resonators. *IEEE Photonics Technology Letters, 19*(10), 741–743.
16. Karplus, M., & Porter, R. N., (1970). Atoms and molecules; an introduction for students of physical chemistry. *Atoms and Molecules; An Introduction for Students of Physical Chemistry*, 1–20.
17. Koch, T. L., Chiu, L. C., & Yariv, A., (1982). Analysis and performance of a picosecond dye laser amplifier chain. *Journal of Applied Physics, 53*(9), 6047–6059.
18. Lawrence, J. R., Turnbull, G. A., & Samuel, I. D. W., (2002). Broadband optical amplifier based on a conjugated polymer. *Applied Physics Letters, 80*(17), 3036–3038.
19. McGehee, M. D., & Heeger, A. J., (2000). Semiconducting (conjugated) polymers as materials for solid-state lasers. *Advanced Materials, 12*(22), 1655–1668.
20. Monroy, I. T., Vd Boom, H. P. A., Koonen, A. M. J., Khoe, G. D., Watanabe, Y., Koike, Y., & Ishigure, T., (2003). Data transmission over polymer optical fibers. *Optical Fiber Technology, 9*(3), 159–171.

21. Moses, D., (1992). High quantum efficiency luminescence from a conducting polymer in solution: A novel polymer laser dye. *Applied Physics Letters, 60*(26), 3215–3216.
22. Nie, J., Lindén, L. Å., Rabek, J. F., Fouassier, J. P., Morlet-Savary, F., Scigalski, F., & Andrzejewska, E., (1998). A reappraisal of the photopolymerization kinetics of triethyleneglycol dimethacrylate initiated by camphorquinone-N, N-dimethyl-p-toluidine for dental purposes. *Acta Polymerica, 49*(4), 145–161.
23. Peierls, R., & Peierls, R. E., (1955). *Quantum Theory of Solids*. Oxford University Press.
24. Pope, M., & Swenberg, C. E., (1999). *Electronic Processes in Organic Crystals and Polymers* (Vol. 56). Oxford University Press on Demand.
25. Rabe, T., Gerlach, K., Riedl, T., Johannes, H. H., Kowalsky, W., Niederhofer, J., & Scherf, U., (2006). Quasi-continuous-wave operation of an organic thin-film distributed feedback laser. *Applied Physics Letters, 89*(8), 081115.
26. Riechel, S., Lemmer, U., Feldmann, J., Berleb, S., Mückl, A. G., Brütting, W., & Wittwer, V., (2001). Very compact tunable solid-state laser utilizing a thin-film organic semiconductor. *Optics Letters, 26*(9), 593–595.
27. Riedl, Rabe, T., Johannes, H. H., Kowalsky, W., Wang, J., Weimann, T., & Scherf, U., (2006). Tunable organic thin-film laser pumped by an inorganic violet diode laser. *Applied Physics Letters, 88*(24), 241116.
28. Sakata, H., & Takeuchi, H., (2008). Diode-pumped polymeric dye lasers operating at a pump power level of 10 mW. *Applied Physics Letters, 92*(11), 106.
29. Samuel, I. D. W., & Turnbull, G. A., (2007). Organic semiconductor lasers. *Chemical Reviews, 107*(4), 1272–1295.
30. Scherf, U., Riechel, S., Lemmer, U., & Mahrt, R. F., (2001). Conjugated polymers: Lasing and stimulated emission. *Current Opinion in Solid State and Materials Science, 5*(2, 3), 143–154.
31. Schwartz, B. J., Hide, F., Andersson, M. R., & Heeger, A. J., (1997). Ultrafast studies of stimulated emission and gain in solid films of conjugated polymers. *Chemical Physics Letters, 265*(3–5), 327–333.
32. Spanggaard, H., & Krebs, F. C., (2004). A brief history of the development of organic and polymeric photovoltaics. *Solar Energy Materials and Solar Cells, 83*(2, 3), 125–146.

33. Tang, C. W., & VanSlyke, S. A., (1987). Organic electroluminescent diodes. *Applied Physics Letters, 51*(12), 913–915.
34. Tessler, N., Medvedev, V., Kazes, M., Kan, S., & Banin, U., (2002). Efficient near-infrared polymer nanocrystal light-emitting diodes. *Science, 295*(5559), 1506–1508.
35. Turnbull, G. A., Andrew, P., Barnes, W. L., & Samuel, I. D. W., (2003). Operating characteristics of a semiconducting polymer laser pumped by a microchip laser. *Applied Physics Letters, 82*(3), 313–315.
36. Vasdekis, A. E., Tsiminis, G., Ribierre, J. C., O'Faolain, L., Krauss, T. F., Turnbull, G. A., & Samuel, I. D. W., (2006). Diode pumped distributed Bragg reflector lasers based on a dye-to-polymer energy transfer blend. *Optics Express, 14*(20), 9211–9216.
37. Vasdekis, A. E., Turnbull, G. A., Samuel, I. D. W., Andrew, P., & Barnes, W. L., (2005). Low threshold edge emitting polymer distributed feedback laser based on a square lattice. *Applied Physics Letters, 86*(16), 161102.
38. Xia, R., Heliotis, G., & Bradley, D. D., (2004). Semiconducting polyfluorenes as materials for solid-state polymer lasers across the visible spectrum. *Synthetic Metals, 140*(2, 3), 117–120.
39. Xia, R., Heliotis, G., Hou, Y., & Bradley, D. D., (2003). Fluorene-based conjugated polymer optical gain media. *Organic Electronics, 4*(2, 3), 165–177.
40. Xia, R., Heliotis, G., Stavrinou, P. N., & Bradley, D. D. C., (2005). Polyfluorene distributed feedback lasers operating in the green-yellow spectral region. *Applied Physics Letters, 87*(3), 031104.
41. Yan, M., Rothberg, L. J., Papadimitrakopoulos, F., Galvin, M. E., & Miller, T. M., (1994). Spatially indirect excitons as primary photoexcitations in conjugated polymers. *Physical Review Letters, 72*(7), 1104.
42. Yan, M., Rothberg, L., Hsieh, B. R., & Alfano, R. R., (1994). Exciton formation and decay dynamics in electroluminescent polymers observed by subpicosecond stimulated emission. *Physical Review B, 49*(14), 9419.
43. Yang, Y., Turnbull, G. A., & Samuel, I. D. W., (2008). Hybrid optoelectronics: A polymer laser pumped by a nitride light-emitting diode. *Applied Physics Letters, 92*(16), 150.

44. Yap, B. K., Xia, R., Campoy-Quiles, M., Stavrinou, P. N., & Bradley, D. D., (2008). Simultaneous optimization of charge-carrier mobility and optical gain in semiconducting polymer films. *Nature Materials, 7*(5), 376–380.
45. Zavelani-Rossi, M., Perissinotto, S., Lanzani, G., Salerno, M., & Gigli, G., (2006). Laser dynamics in organic distributed feedback lasers. *Applied Physics Letters, 89*(18), 181105.
46. Zhang, Z. M., (2000). Surface temperature measurement using optical techniques. *Annual Review of Heat Transfer, 11*.

CHAPTER 6

ORGANIC SEMICONDUCTORS FOR PHOTODETECTORS

CONTENTS

6.1. INTRODUCTION

Photodetection has become a leading technique in the development of science and technology as a result of our advancement into the information era. In photodetection apps, photodetectors that transform occurrence photons into an electric signal are widely employed. Owing to its high charge-carrier mobility, low exciton bond length, and excellent stability, the majority of commercial photodetectors utilize photodiodes made from semiconductor materials, like silicon, or substances, like III-V semiconductors (Newman et al., 2004; Brütting, 2005). The new uses of advertising inorganic photodetectors (OPDs) are severely constrained, particularly in light of the requirement for large-area, adaptable, and low-cost gadgets, owing to several disadvantages including expensive and complex production processes and physical lack of flexibility (Takimiya et al., 2011).

Organic solar cells are natural matter with special properties, like electrical properties that fall among that of steel and a dielectric material. Their constituents are–bonded compounds or polymers composed of hydrogen and carbon atoms (often with heteroatoms like nitrogen, sulfur, as well as oxygen). Typically, small organic semiconductors (OSC) are produced by a deposition technique, whilst polymeric semiconductors are handled in solution by dip covering plus able to print. Even though OSC are also recognized as semiconductors, they vary away from conventional inorganic semiconductors (Yamashita, 2009; Myers and Xue, 2012).

OSC have weak van der Waals forces among particles, although inorganic semiconductors are made up of covalent bonding among two atoms. The compositional and structural distinctions between inorganic and OSC result in major distinctions in their structural, electronic, and refractive indices.

In physical and mechanical properties, the weak van der Waals relationships among separate molecules create OSC usually softer than semiconductor materials constituted of strong covalent bonds, like Si and Ge. The covalent as well as the periodic configuration of inorganic semiconductors findings in the transfer of electrons and power bands for electrochemical characteristics (conduction band as well as valence band). Even so, OSC have an energy gap rather than bands (the lowest unoccupied molecular orbital (LUMO) and the highest occupied molecular orbital (HOMO) regarding the end of valence as well as conduction bands). Owing to the excessive bandgap energy in OSC, thermal activation produces just several free means of transport. On such doping or photoexcitation, the suitable electrode injects free charges (Knupfer, 2003; Köhler and Bässler, 2009).

Furthermore, the actions of load transfer in OSC are likely to depend on molecular hopping, which typically results in lower movement in OSC compared to inorganic semiconductors. In terms of optoelectronic characteristics, OSC are more absorbent than their inorganic counterparts. In this regard, OSC are ideally suited for the manufacture of optoelectronic gadgets like solar cells as well as photodetectors. Owing to its low dielectric permittivity (static dielectric constant, 2–4) (Knupfer, 2003), their main optical excitations are balanced excitons with classic Coulomb-binding energy of 0.1–2.0 eV; as an outcome, it is hard to distinguish such charges. According to this viewpoint, OSC are not appropriate for optoelectronic gadgets till suitable donor/acceptor connections are developed to address the issue (Dong and Hu, 2010; Jacob, 2014).

Kudo and Moriizumi first invented OPDs fabricated with the organic small-molecule dyes merocyanine (MC) as well as rhodamine B (RB) exhibiting a visible (vis) spectrum reply variety in 1981.

In a vacuum, such dyes were sublimated to create thin films. The thorough framework of the gadget was ITO/ZnO/MC/RB/Al. Even though unconfirmed, the authors acknowledged the significance of the MC/RB interaction for photocatalytic production. Tang reported organic thin-film photovoltaic cells along planar interfacial donor-acceptor heterojunction and 1% conversion efficiency under AM2 light in 1986 (Knupfer, 2003). In 1992, electron transfer between a conjugated polymer and fullerene derivatives was independently demonstrated by Sariciftci et al. and Morita et al The important part of donor-acceptor phase detachment in accomplishing appropriate charge carrier pathways for trying to collect electron-hole pairs has been identified. Sariciftci et al. published the initial presentation of planar heterojunction polymer in 1993 photodiodes and photovoltaics 1995 also saw the development of bulk heterojunction solar cells and photodiodes with polymer–fullerene, and polymer-polymer blends (Krause et al., 2008; Lin et al., 2012). After that, the development of small-molecule, as well as polymer OPDs, accelerated. From a structural standpoint, OPDs can be categorized as organic photodiodes, organic photoconductors, and organic phototransistors, in addition to their classification according to the type of material. Both photodiodes and photoconductors are two-terminal machines, while phototransistors have three. Now, we'll focus primarily on the evolution of OPDs with two terminals. Organic phototransistors-focused readers are referred to additional analysis articles (Nakano et al., 2015; Hofmann et al., 2019).

The study on OPDs has made big progress over the years, and their achievement has managed to improve to the point where some are now significant compared to those of inorganic photodetectors. There have also been several accomplishments in application-oriented research. This evaluation begins with an explanation of the fundamental principle of operation of OPDs, followed by a discussion of the current progress of performance analysis and likely new implementations. Eventually, we investigate the overall development of OPDs (Torabi et al., 2015; Schwarze et al., 2016).

6.2. WORKING PRINCIPLE OF ORGANIC PHOTODETECTORS (OPDS)

The operating principle of OPDs, like that of inorganic photodetectors, is associated with the internal piezoelectric effect. Figure 6.1 depicts the working principle of OPDs, which comprises four stages for converting an optical signal to an electrical signal: (i) the production of excitons as a result of absorbing light; (ii) the dispersion of the produced excitons to the donor/acceptor interaction to generate charge carrier excitons; (iii) the isolation of charge transport excitons into the free charge carriers just at donor/acceptor interaction; and (iv) the selection of free charge carriers at the diodes (Shen et al., 2004; Meng et al., 2009).

An exciton is a connected state comprising an electron as well as a hole influenced by one another by the electrostatic Coulomb interaction. Photon absorption in a transistor or the interest among holes and electrons infused from diodes can both produce this condition. Excitons are divided into three categories based on their size compared to the interparticle range: Frenkel excitons, charge transport excitons, and Wannier-Mott excitons. Because of the weak intermolecular van der Waals interactions, particles in organic solids are more separated apart than in covalent compounds. Organic solids have more concentrated charge carriers than inorganic solids. Whenever an organic material soluble a photon, it produces a Frenkel exciton confined at one particle or a charge transport exciton with enhanced delocalization, with a radial distance of two or more lattice factors. Wannier-Mott excitons, which occur in covalent intrinsic semiconductors, are uncommon in OSC (Zhang et al., 2015; Root et al., 2016).

Frenkel excitons, that have a good bond energy, are frequent in OSC due to their low dielectric permittivity. The separation of an exciton from such a big bond length into charge transport needs tens of meV of input energy, whereas crystalline semiconductors need just several meV since they are less tightly bound.

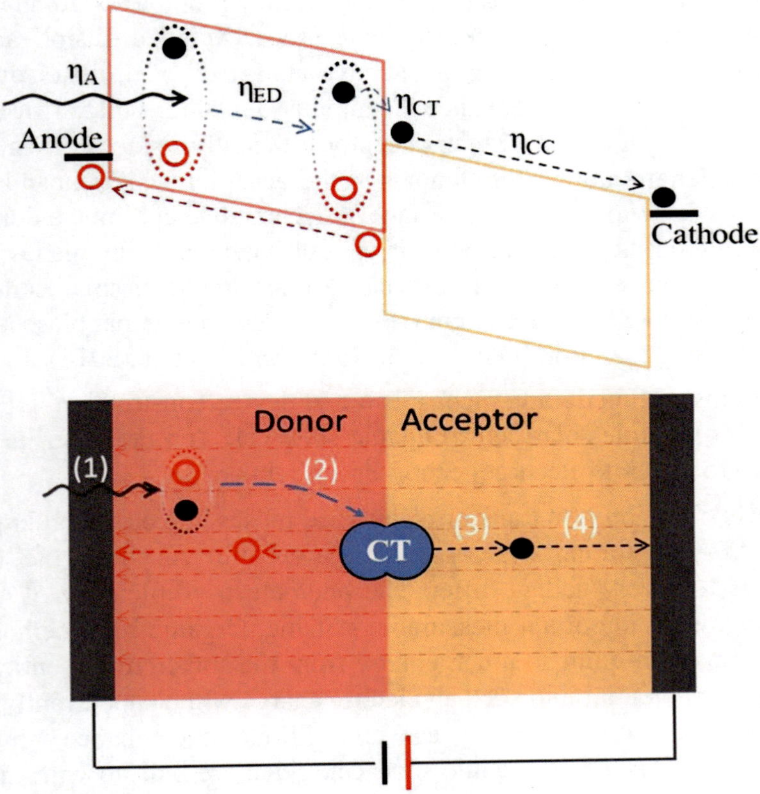

Figure 6.1. Working principle of OPDs.

Source: https://www.researchgate.net/figure/a-Working-principle-diagram-of-inverted-OPDs-b-Energy-level-diagram-of-devices_fig5_356107136.

Organic semiconductor excitons need not naturally dissolve into free electron-hole couples. While electric fields may be used to aid in the method of exciton parting, the electric field required for exciton separation could be as strong as 106 V cm, putting the method unsuitable for applied uses

(Tokumoto et al., 1987; Henson et al., 2012). In low electric fields, though, an effective exciton process happens at the donor-acceptor contacts. As a result, the excitons generated in organic molecules by absorbing light are only helpful whenever they are distributed to the donor-acceptor interaction. As a result, the length of the exciton diffusion is a key component in calculating the frequency that perhaps the exciton will approach the donor-acceptor contact, where an ideal diffusion length (Ld) may promote exciton breakup. Regrettably, most organic molecules have a short exciton diffusion range. To address this issue, substances going to have a long exciton diffusion distance, like C_{60}, are often used to enhance the number of excitons that diffuse to the contact. Mixing a donor with an acceptor efficiently reduces the means of molecular length among the donor and the acceptor while expanding the surface of the donor/acceptor interface. On average, combining a donor and an acceptor limits the movement of the substance and diminishes charge carrier effectiveness, which is especially undesirable in solar cells. This phenomenon may be decreased in OPDs by using an external biased, which aids in charge separation (Pecile et al., 1989; Martín et al., 2018).

In terms of functioning mechanism and device construction, OPDs are identical to organic solar cells. Organic solar cells, on the other hand, are designed to generate electrical energy from light energy.

OPDs' function is to transform electrical pulses into electrical impulses. Since every operating system serves a specific function, the quality standards for every gadget differ. The photoelectric efficiency of organic solar cells is an important measurable statistic. Organic solar cells should indeed transform quite so much energy from the sun from the entire solar spectrum as potential into such an electric charge while concurrently trying to balance the amount of voltage as well as fill factor to enhance conversion efficiency. In contrast to organic solar cells, that are built to withstand the entire solar spectral range, the spectral variety of OPDs must be optimized for the particular application. Other quality factors of OPDs, like specific detectivity as well as response speed, are especially important regarding spectral data. Furthermore, applying a negative voltage to photodetectors can improve their efficiency (Zhu et al., 2008; Jiang et al., 2013).

As a result, photodetectors can operate in the 3rd quatrain or the negative aspect of the Y-axis, whereas organic solar cells could only operate in the 4th quadrant (Figure 6.2). Figure 6.3 depicts a circuit diagram of an organic photodiode. Whenever the gadget has functioned under fault current situations (load resistance RL 0), the brief circuit current (Isc) is the current iteration.

Whenever the machine has functioned under open circuit conditions (RL), the open circuit voltage (Voc) is the output signal. The shunt resistance is the voltage-to-current ratio near 0 V. The terminal capacitance (Ct) is the gadget's electric potential that is connected to the gate voltage. As illustrated in Figure 6.2, both Voc as well as Isc rise as the brightness of the light that enlightens the photodetector.

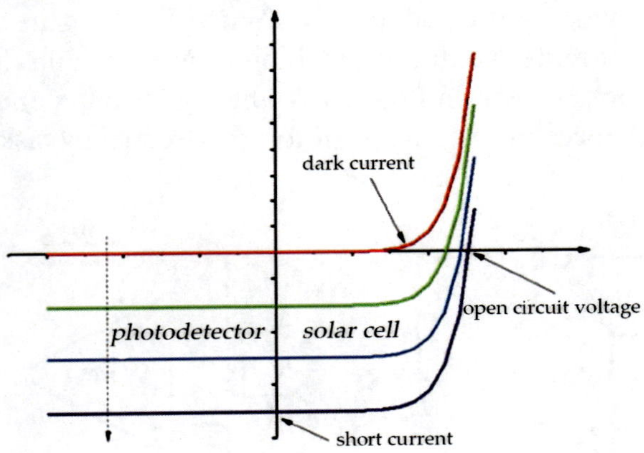

Figure 6.2. In a Cartesian coordinate system, the value of the current as well as the voltage of organic photodiodes as well as organic solar cells.

Source: https://onlinelibrary.wiley.com/doi/abs/10.1002/adom.201800522.

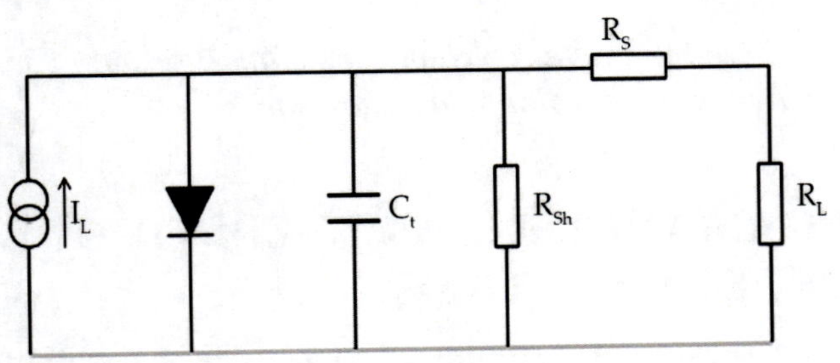

Figure 6.3. A photodetector's equivalent resistance.

Source: https://electronics.stackexchange.com/questions/509925/photodiode-equivalent-circuit.

In theory, both current and voltage may be employed as photodetector output signals. Current is often used since it has superior uniformity and reaction speed. Usually, the modest output photocurrent is increased for additional detection.

Photodetectors may work for either photoconductive or photovoltaic modes. The photodiode is short-circuited in the photoconductive state. OPDs measure the change and uniformity may be considerably enhanced by using negative voltage, which is comparable to traditional inorganic silicon semiconductor photodiodes. The negative bias, on the other hand, enhances the current. The photodiode is neutral biased in photovoltaic operation. Such a way of operation utilizes the photoelectric effect, which would be the foundation of solar cells. When working in a photovoltaic state, the photocurrent is negligible, making it suited for ultralow-light-level uses. Furthermore, as seen in Figure 6.4, the photovoltaic mode provides a straightforward operating arrangement for low-frequency tasks (Newman et al., 2004; Zhang et al., 2017).

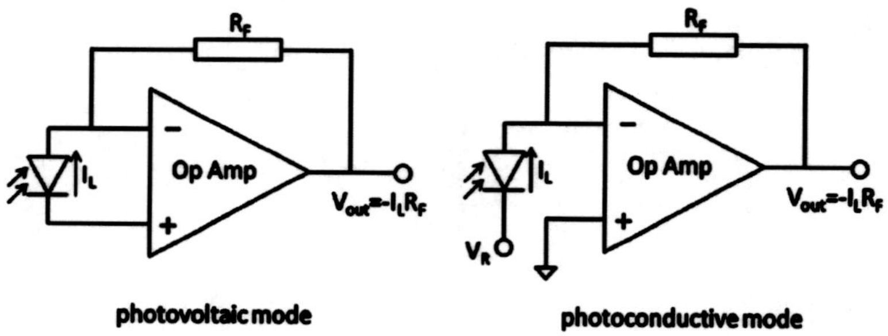

Figure 6.4. Photodetector amplifier circuits in photovoltaic as well as photoconductive modes.

Source: https://electronics.stackexchange.com/questions/404335/photodiode-in-photoconductive-vs-photovolatic-configuration.

6.3. PERFORMANCE PARAMETERS OF ORGANIC PHOTODETECTORS (OPDS)

A photodetector's external quantum efficiency (EQE), photoresponsivity, dark current, detection sensitivity, spectra responses, uniformity, as well as the speed of response are all great factors of performance. It is significant to remember that the importance of these factors changes based on the needs of the application, resulting in varied device designs. In this part, we describe these factors and their relative development (Léonard and Tersoff, 2002; Chang et al., 2004).

6.3.1. EQE and Responsivity

An essential measure describing the photoelectrochemical capabilities of an OPD, the EQE is the relation of the average amount of photoelectrons created per time unit to the amount of instance photons at a certain wavelength. The EQE of OPDs is normally assessed utilizing a phase-sensitive lock-in approach, which eliminates the impact of current on the instrument. The EQE reflects the ratio among recovered electrons and incoming photons.

$$EQE = \frac{I_p / e}{p / hv}[\%]$$

wherein; I_P is the observable photocurrent; *e* is the element charge; *P* is the irradiation optical power density; *h* is Planck's constant, as well as is the light's vibrational speed (Zhao and Mazumdar, 2004).

Photoresponsivity (R) is the ratio of photons to incident light power at a certain wavelength. Thus, the response is measured as A W1 (as the ratio of current to power). The responsibility may be stated as follows:

$$R_\lambda = \frac{J_{photo}}{P} = \frac{J_{light} - J_{dark}}{P}$$

where; J_{photo} is the photocurrent concentration; J_{light} is the illumination current; J_{dark} is the dark current; and *P* is the incoming light's input optical concentration. Consequently, the EQE and *R* at a particular wavelength have had the following expression, λ.

$$EQE = \frac{R \times 1240}{\lambda} \times 100[\%]$$

The EQE is among the most important performance attributes in photodetectors, and it is often optimized for OPDs. Figure 6.4 depicts the processes capable of transitioning optical pulses to an electrical charge signal. The EQE of OPDs is dependent on various four efficiencies:

$$\eta_{EQE} = \eta_A \eta_{ED} \eta_{CT} \eta_{CC}$$

A represents the light absorption efficiency, ED represents the exciton diffusion efficiency, CT represents the dissolution efficiency of charge transport excitons, and CC represents the excess charge carrier efficiency (Wang et al., 2005; Zou and Zhang, 2012).

The value of A is dependent on the film's organic components. The absorbing spectrum coefficients of organic compounds are generally higher (105 cm^{-1}) than either inorganic materials (103 cm^{-1}). Consequently, organic material coatings with such a depth of many nanometers fulfill the criteria for absorption. Furthermore, because obtained using the proposed organic compounds is ruled by thermally induced hopping among extremely concentrated sites (or traps), the movement of organic compounds is considerably lower as compared to inorganic compounds, which indicates that the maximum depth of the organic layers must not be excessively thick so as not to reduce the efficiency of the charge carrier. Consequently, the overall thickness of OPDs generally varies from hundreds of nanometers to less than one micrometer, on a similar scale as the wavelength of light within the organic layers. Concerning the double layered panel of an organic phase engrained among a transparent conducting and a reflective metal diode, which is common in OPDs, the similar wave train of the light event on the transparent electrode as well as light reflected by the metal electrode converges in the active layer. In this instance, optical interference happens in the active layer, and the strength value of the light source across the organic layers is non-uniform. Since light absorption is dependent on the strength of the light source, it is vital to optimize the dispersion of the light field by optimizing the specifications of every layer to improve the optical performance (Zou and Zhang, 2012).

Because the absorption of light depends on the intensity of the light field, the optimization of the design parameters of each layer is necessary to tune the distribution of the light field, to maximize the optical absorption, or meet the particular absorption requirement for practical applications. The well-established transfer matrix technique has been applied to calculate the optical field in organic thin films,which are schematically shown in Figure 6.5

.(Stewart and Léonard, 2004; Marcus et al., 2006).

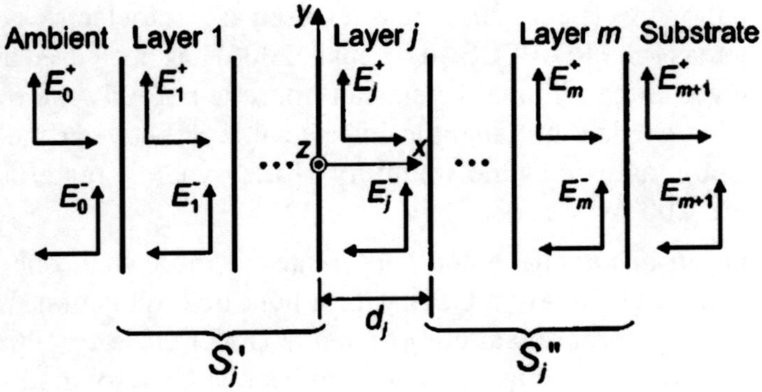

Figure 6.5. A generic multilayer construction with m layers positioned among a semi-infinite translucent ambiance and a semi-infinite substrate. Every layer j (j = 1, 2, …, m) has a depth (d_j), and its optical qualities are characterized by complicated refractive index. The optical electric field at every location inside a layer consists of two main parts: one traveling in the positive x-direction as well as the other in the negative x-direction.

Source: https://www.researchgate.net/publication/234946771_Modeling_phoocurrent_action_spectra_of_photovoltaic_devices_based_on_organic_thin_films.

Since the bonding energy of photogenerated excitons in OSC is rather high, they are not easily dissolved into free charge carriers by a modest electric field. Typically, an exciton is transformed into a charge transport exciton, which is more readily dissolved into a free charge couple at the donor-acceptor contact. The excitons must propagate to the donor/acceptor contact for bilayer-structured OPDs. Given the short diffusion length of excitation energy in semiconductor materials, some excitons are lost before hitting the donor/acceptor contact. To prevent such energy loss, the optically active surface of biological bulk heterojunction photodiodes is often a mixture of donor-acceptor elements. Following charge separation, photogenerated holes, as well as electrons, should be delivered to the indicator electrode, accordingly, via the donor as well as acceptor material mixture. Consequently, the movement of the composite film is lesser than those of the separate acceptor or donor substance, which limits the transfer rate of charge transport. Unlike organic solar cells, OPDs may be made more efficient by introducing a reverse bias. The reverse bias improves the EQE of OPDs by a large margin. In addition, optimizing the electron/

hole removal levels can enhance the EQE of the manufactured OPDs. For example, the pass conjugated polyfluorene, TiO_2-C_{60} bulk composites, and zinc oxide nanowires (Lee, 2007) are often used as photoelectric conversion surface, whereas PS-TPD-PFCB (Itkis et al., 2006) has been used as the hole extraction layer. Such removal layers can improve not only the removal of photogenerated carriers but also the interfacial area between the donor as well as acceptor, as well as the solubility of the dynamic material (Itkis et al., 2006; Lee, 2007).

In broad, an absorbed photon can create no more than one electron-hole couple in an OPD, so an OPD with a bulk heterojunction framework has an optimum internal quantum efficiency (IQE) close to 100%. In the amorphous system of donor-acceptor molecular heterojunctions, prior research suggests that the EQE reaches 100%. Regular devices cannot, nevertheless, achieve an IQE greater than 100% due to inevitable energy losses. Thru the structural selection of material, this circumstance can be modified in OPDs. The singlet fission reaction in pentacene is capable of converting a single photogenerated singlet excited state into two triplet excitons, thereby possibly twice the quantum efficiency of OPDs. Lee et al. improved the EQE of an organic layered photodetector by using the exciton fission impact of pentacene. As seen in Figure 6.6, the energy of the singlet exciton is larger than two times that of the very first three same excited states in pentacene. Consequently, the transformation of a single exciton into two triplets is actively feasible in the spin-allowed state of pentacene. The writers created a multilayer photodetector representing the different ultrathin donor as well as acceptor levels of pentacene and C_{60}, that enabled effective exciton separation and charge extraction. Thus, the EQE was greater than 100% at 660 nm. In fact, in addition to exciton fission, the EQE of an organic photodetector can be greater than 100% by employing an advanced structure in which charges could be magnified for injection into the gadgets. As the operating method of gain-type OPDs differs from that of photovoltaic devices, we give an isolated overview here (Freitag et al., 2004; Junaid et al., 2020).

6.3.2. Dark Current, Noise, and Sensitivity

Another key consideration for photodetectors is the dark current (or leakage current), which is a weak signal current flowing thru gadgets in the dark underneath a reverse voltage. The vibration in photodetectors is primarily caused by the conversion of input light into current, as well as the participation of innate noise such as random fluctuations, shot noise, as well as impulse noise.

Thermal noise (also known as Johnson noise or Nyquist noise) is the automated noise produced by the thermal agitation of electric charge (generally electrons) within an electrically conducting at balance, which occurs irrespective of the applied voltage. Thermal noise in a photodetector is caused by carrier thermalization and is proportional to parallel resistance as well as temperature. In denotes the magnitude of random fluctuations (Chen et al., 2005; Marty et al., 2006).

$$I_{jn} = \sqrt{\frac{4k_B T \Delta f}{R_{SH}}}$$

where; k_B is the Boltzmann constant; T is the temperature in degrees Kelvin; f is the noise level bandwidth, and RSH is the gadget's shunt resistance. Shot noise is a form of electronic sound that could be represented mathematically as a Poisson process. The specific nature of electrostatic force causes shot vibration in electronics (Pop et al., 2005; Phillips et al., 2006). The numerical variations of both the photons and the dark current contribute to shot noise in photodetectors. The amplitude of the shot noise I_{sn} is given below:

$$I_{sn} = \sqrt{2q(I_P + I_D)\Delta f}$$

wherein; $q = 1.6 \times 10^{19}$ C is the electron charge; I_P is the photogenerated load; I_D is the dark current; and f is the noise level bandwidth. Whenever a photodetector is operating in a photoconductive state, shot noise is often regarded as the major source.

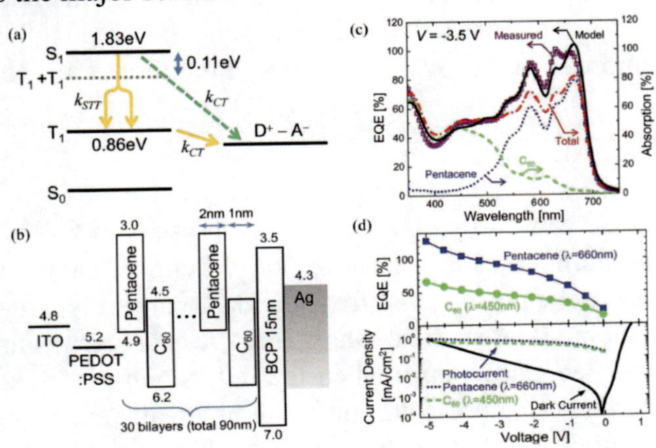

Figure 6.6. (a) Transfer of energy methods in a photodetector based on pentacene/C_{60}. A singlet exciton is followed by two triplet excitons. (b) An energy-level diagram of a multiple layer pentacene/C_{60} photodetector. (c) An EQE

spectra as well as the adsorption process of a pentacene/C_{60} multilayer gadget at V = 3.5 V. (d) The EQE in the dark and under irradiation as a function of the applied voltage and current-voltage parameters.

Source: https://aip.scitation.org/doi/10.1063/1.3182787.

Flicker sound is a low-frequency sound having an inverse proportion of noise level to frequency. Flicker noise has been found in OPDs, according to research. The sources of noise mentioned above are distinct. As a result, an equation may be used to define the overall noise (Wei et al., 2004):

$$I_n \aleph \sqrt{I_{sn}^{\text{ü}} \quad I_{jn} \quad I_{fn}}$$

The lower bound of light energy recognition for a photodetector is often defined as the incoming light power necessary to create a signaling current equivalent to the noisy current. The noise equivalent power (NEP) of a photodetector is an important quality parameter, described as the optic power input providing a signal-to-noise ratio of one.

$$NEP = \frac{I_n}{R}$$

The detecting capability is represented by Detective (D), the reciprocal of NEP. Even though the noise level is connected to photodetector effective area as well as bandwidth, specific detectivity (D*), which is acquired by equating D concerning device area (A) and bandwidth (f), is being used to make a comparison of detection capabilities among different photodetectors. D* is measured in cm Hz12 W1 units (or Jones). A higher D* indicates performance improvements (Wei et al., 2004; Simmons et al., 2007).

$$D^* = \frac{\sqrt{A\Delta f}}{NEP} = \frac{R\sqrt{A\Delta f}}{I_n}$$

Thermal noise caused by shunt resistance persists in OPDs regardless of whether an external bias is present or not. Whenever the photodetectors are reverse biased, the noise signal from the dark current is bigger than the random fluctuations. Real statistics show that flicker noise occurs in OPDs at a lower frequency, particularly when the reverse bias is high. At higher frequencies, the flickering noise diminishes dramatically and approaches the estimated shot sound level. The shot noise resulting from the dark current is often assumed to be a significant element of the total noisy flow. As a result, special emphasis has been made on decreasing the dark current density to decrease shot disturbance, and the cause of flicker noise has not been well known. Using ultrathin alternated donor-acceptor films in a stack configuration for OPDs based on tiny molecules, Xue and Forrest effectively regulated the dark current in reverse bias. The electron/hole filtering layers are often employed in OPDs to minimize dark current and prevent charge-carrier insertion under reverse bias. Although the EQE decreases with the decrease of dark current via this approach, the computed D* increases owing to the greater drop in dark current relative to the EQE. This approach works especially well in OPDs with a near-infrared (NIR) wavelength sensitivity (Hecht et al., 2006).

Across the whole spectrum range, the specific detectivity was capable of reaching 1,013 Jones. Liu et al. disclosed a NIR organic photodetector with a way to solve reversed structure as well as electron and hole extraction layers of ZnOx and MoO_3, correspondingly. The photocurrent was enhanced while the dark state current was decreased, resulting in a gadget with a specific monitor utilizing more than 1,012 Jones in the 400 to 850 nm wavelength range. Armin et al. created polymer photodetectors having reactive layer thicknesses ranging from 100 to 700 nm using a PCDTBT and PC71BM mix. The authors discovered that a dense energetic layer assisted in lowering the dark current and maintaining a high EQE. In OPDs using F8BT: PDI as the active environment, the temperature dependency of the dark current was discovered (Larrimore et al., 2006).

To describe the phenomena, a two-stage mechanism comprising of a temperature-independent phase for electric-field-assisted intercalation from the connections to the active material and a substantially triggered phase for transmission all over probes was suggested. At this point, the dark current density of OPDs has indeed been lowered to several nA cm^2 or even lower, approaching the standard of established silicon technology and being especially beneficial for large-area detection. The theoretical estimation of shot noise obtained from dark current, on the other hand, may overstate the D*. Using a novel low band-gap polymer, Yao et al. investigated the noise properties of an organic NIR photodetector. Figure 6.7 shows the emergence of 1/f noise whenever the NIR photodetector was subjected to high reverse voltages. The noise current that's within 3 dB of the image noise limit was at a resolution of 4 kHz. Kim et al. got comparable findings while studying the noise spectrum efficiency of a solution-processed nonpolymeric organic photodetector. The flicker noise was inversely correlated to frequency (1/f) for frequencies less than 10 Hz. The flicker noise fell dramatically at higher frequencies, over 10 Hz, and achieved a stable value that was near to the estimated shot noise. Both of those investigations demonstrated that the NEP in OPDs at high frequencies is primarily restricted by shot noise. Several additional investigations have presented the presence of flickering noise at a lower frequency (Loucif-Saibi et al., 1993).

Xiong et al. recently discovered the lack of 1/f noise in OPDs from 5 to 50 Hz by transfer-printing poly-hexylthiophene (P3HT) films as the electron-blocking coating. Nevertheless, the cause of flicker noise at lower frequencies was previously unknown. As a result, more research should be conducted to minimize or remove 1/f noise.

6.4. SPECTRAL RESPONSE CHARACTERISTICS

Photodetectors' light sensitivity changes with wavelength. The spectral response parameters are the relationships among photosensitivity as well as wavelength. Currently, silicon photodetectors with a spectral response span of 200–1,100 nm are often used in industrial cameras. To remove interference from the Sun's NIR radiation, infrared filters are therefore necessary. The most notable benefit of OPDs over inorganic photodetectors is the ability to customize the electronic and optical characteristics of the used organic compounds to alter the spectral variety from ultraviolet (UV)-Vis to NIR wavelengths. Below, we go through the OPDs associated with various spectral ranges (Valentini et al., 2006).

6.4.1. Visible Organic Photodetectors (OPDs)

Electromagnetic radiation having wavelengths between 380 and 780 nm is considered a spectrum because it can be seen by the naked eye. The camera is significant to use of noticeable photodetectors. The majority of cameras now available are built using inorganic silicon semiconductors. The two primary techniques for the imaging system in a camera are complementary metal-oxide semiconductor (CMOS) and charge-coupled device (CCD). Due to the "color blindness" of CMOS and CCD photodiodes, filtration placed at the front of the camera enables the devices to distinguish between red, green, as well as blue pixel resolution. An infrared cutoff filter is added to such filters to avoid infrared radiation that reaches the sensors in real-world color picture operations (Guldi et al., 2005).

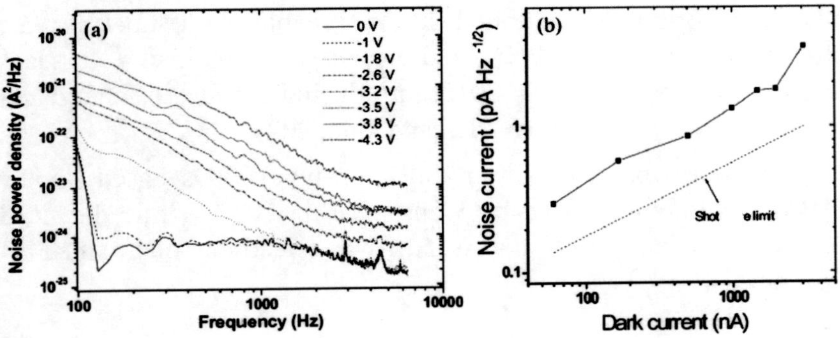

Figure 6.7. (a) Evaluated dark current acoustic spectra at various biases. (b) A performance of the evaluated noise current and dark current.

Source: https://pubs.acs.org/doi/10.1021/acsomega.0c04136.

This necessity not only makes setup more expensive and difficult, but it also necessarily lowers the photoresponse of equipment. Therefore, for colored picture sensing, multicolor, unfiltered visible-light photodetectors with such a narrowband reaction would be helpful. The majority of OSC have high light absorptivity for visible light. To suit the requirements of narrowband light sensing, the absorption spectra band may be modified, similar to how OSC' chemicals required to transmit can be changed. It has been made possible to create OPDs lacking color filters that solely react to blue, green, and red light. OPDs' spectrum response range may also be modified by using optical phenomena such as surface plasmons (SPs), the interfering impact, and the microcavity impact.

In addition to the previously described two techniques, OPDs includes a built-in filter. It has been effectively established that red-light OPDs containing an excessive number of organic layers of,-diphenyl sexithiophene (P6T) and BP3T may lessen the blue-light sensitivity of OPDs built on copper-phthalocyanine (CuPc)/C_{60}. All P6T and BP3T are thiophene-based compounds with strong hole-transporting characteristics, as demonstrated in Figure 6.8. The blue light was absorbed by the 100 nm P6T layer, preventing it from accessing the photoactive surface, while the wide band gap of the BP3T layer stopped CuPc from receiving an excited state from P6T. The CuPc/C_{60}-based OPDs' responsiveness to blue light was greatly lowered by the mixture of the light-absorbing barrier as well as the exciton-blocking layer, while the redlight sensitivity was not adversely affected. OPDs with a restricted spectral response has also received more focus in recent times. Burn's Group originally exhibited the first visible-blind sub-100 nm full-width-at-half-maximum (FWHM) red as well as NIR photodetectors lacking the usage of input filtration in 2015, introducing the idea of charge carrier shortening (Cai et al., 2017; Herrick and Guo, 2021).

Highly narrowband photomultiplication-type OPDs based on P3HT: PC71BM were demonstrated by Wang et al. in 2017. While the EQE hit 5,300% with a bias of 60 V, the FWHM was effectively maintained at less than 30 nm.

6.4.2. Ultraviolet (UV) Organic Photodetectors (OPDs)

Electromagnetic radiation having wavelengths among visible light as well as X-rays is known as UV light. UV may be classified into three variations: UVA (400–320 nm), UVB (320–280 nm), and UVC (280–100 nm), as per ISO standard ISO-21348. That after widely utilized laser detection as well as infrared revelation technologies, UV revealing has drawn significant interest in contemporary photoelectric detection technology and has emerged as a significant photodetection technique. Uses for UV detection are many in the industrial, aeronautical, medical, telecom, and military sectors. Semiconductor photodiodes have low-cost manufacture, great linearity as well as sensitivity, and the capacity for high-speed processes, giving them more ideal for UV detection than thermal detectors, photomultipliers, and CCDs. As a result, SiC, Si, ZnO, and III-nitride semiconductor-based inorganic UV photodetectors have received substantial research. These UV photodetectors offer the benefit of versatility, low weight, as well as reasonable. Organic UV photodetectors are an additional option that has far easier production methods. Lin et al. presented organic UV photodetectors employing triarylamine and oligoaryls including oxadiazole as the effective electron transmission donor-acceptor in 2005 (Karplus and Porter, 1970; Pope and Swenberg, 1999). In the UVA, a high EQE and visual blindness were attained.

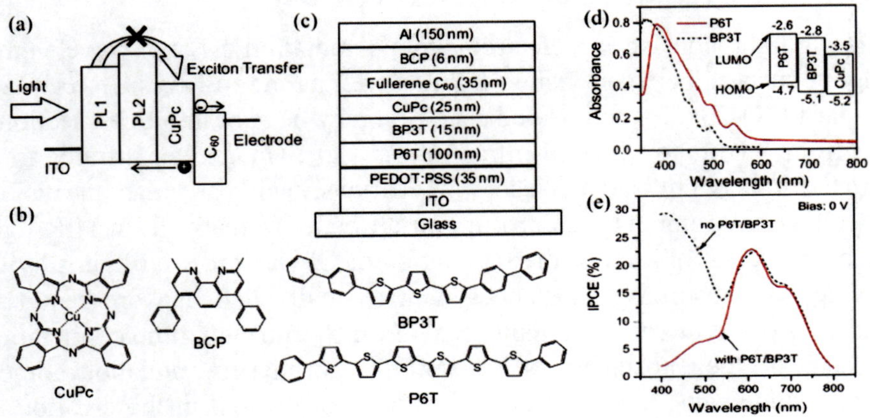

Figure 6.8. (a) shows the energy curve of an organic photodetector built on CuPc/C_{60} that is solely sensitive to red light. (b) The chemical compositions of CuPc, BCP, BP3T, and P6T. Structure of the gadget. (d) The P6T and BP3T thin films' absorption spectra. (e) IPCE spectrum of gadgets with ITO/P6T/BP3T/CuPc/C_{60}/BCP/Al and ITO/CuPc/C_{60}/BCP/Al structural configurations.

Source: https://www.researchgate.net/figure/Electronic-absorption-spectra-of-C-60-in-toluene-curve-1-and-in-n-hexane-curve-2_fig2_257855417.

In 2008, a tris-(8-hydroxyquinoline) gallium (Gaq3) under irradiation of 365 nm UV light was verified (Nie et al., 1998) A thin Al cathode of 12 nm was used as an incident beam slit to establish a high-response OPD with a 280 nm frequency response peak since indium tin oxide (ITO) and glass substrates strongly absorb UV light at wavelengths less than 300 nm (Eastham, 2000). Underneath 280 nm UV light with a strength of 0.428 mW cm^2 at 8 V, a peak photoresponse of 309 mA W1 was attained.

The half-cycle operating duration was 440 min under a 0.18 mW cm^2 280 nm UV light intensity. The UV transparency anode may also be PEDOT: PSS polymer sheets on a quartz substrate as an alternative to the thin Al cathode. Utilizing PEDOT: PSS, Zhu et al. effectively extended the spectrum range of manufactured OPDs to the deep UV area (PH 1000). It has also been possible to create organic UV photodetectors that have noticeable and quasi-solar-blind reactions (Nie et al., 1998; Eastham, 2000).

6.4.3. NIR Organic Photodetectors (OPDs)

Infrared light consists of electromagnetic radiation having wavelengths ranging such as between visible light as well as microwaves. As per ISO standard ISO-20473, the infrared spectrum may be split into three sections: NIR from 780 nm to 3 m, mid-infrared from 3 to 50 m, and far-infrared from 50 m to 1 mm. Night vision, telecommunications, healthcare, machine vision, as well as environmental control might all benefit greatly from NIR-light detecting technologies. For the fabrication of OPDs with a NIR sensitivity, organic NIR-absorbing substances with a small band gap are required. The energy gap in polymers may be reduced by enhancing the conjugation duration of the substance. NIR-absorbing porphyrins, phthalocyanines, heptamethine salts, and squaraine dyes have been used in the creation of NIR OPDs for small molecules and oligomers (Yan et al., 1994; Schwartz et al., 1997).

As a result of their rising use in portable electronics and biological applications, NIR OPDs are appealing prospects for future electrical items. To achieve this aim, elevated NIR OPDs must be manufactured. In this chapter, we offer a general review of improvements in NIR OPDs during the last 10 years from the standpoint of selection of materials as well as gadget optimizations, as well as a summary of the application potential of NIR OPDs. Finally, we have a detailed conversation regarding the problems and opportunities for the further promotion of natural NIR OPDs, Perzon et al. published the low band gap, conjugated polymer LBPP-1 composed of alternating repeats of a dialkoxyphenylene unit and a low-band-gap section. In the section with a smaller band gap, the D–A–D arrangement decreased the bandgap to 1.0 eV. Thus, up to 1,200 nm of light absorption as well as photovoltaic reaction were observed. The discovered NIR-absorbing polymers may react to NIR light with a wavelength of approximately 1,800 nm, which is comparable to those of photodetectors depending on InGaAs at room temperature (Heliotis et al., 2003; Xia et al., 2004).

By radially expanding the coupling of the ¯electron systems, porphyrin tapes may be changed to provide an even higher wavelength sensitivity. As a NIR-absorption substance, Zimmerman et al. created a novel kind of OPD in 2010 by using triply connected porphyrin tape dimers. The scientists discovered that the efficiency of the gadget was affected by functionalization with various side group molecules, which impact the crystal size and morphology of the film. The EQE achieved 6.5%, as well as a fast response of $\tau = 2.12 \quad 0.02$ ns ns, was measured at 1,350 nm. Thru the application of additives, the scientists increased the highest EQE at 1,400 nm to 13.5 0.3%, which was about twice that of earlier NIR OPDs.

Utilizing the direct interparticle charge transport state (CTS) absorption to expand the photo response into the NIR area without the necessity for low bandgap substances is a newly suggested elegant approach (Koch et al., 1982). The CTS is formed by the reaction between the donor's HOMO and the acceptor's LUMO. That phase is only seen at the donor-acceptor interface, where distinct electronic orbital angular momentum combines to generate a new, middle position. Typically, the latter has substantially lower transition energy than that of the elements of a type II heterojunction, like conventional donor/acceptor mixtures. In other respects, the HOMO intensity level of the donor is greater than that of the recipient, but the LUMO residual energy of the acceptor is inferior to the supplier (staggered gap). Because the CTS has lower transition energy, the mixture has the capability of absorbing light at excitation energies below the bandgaps of its components. Due to the intermolecular nature of the CTS state, its absorption intensity is about two orders of magnitude less than that of singlet absorption (Zavelani et al., 2006). To produce a meaningful photoresponse, it is often necessary to use very thin active layers at reverse bias voltages due to the CTS's inherent low excitation energy (Chen et al., 2017). In a conventional device arrangement, increased viscosity of the active region by tens of microns to boost the CTS absorption necessitates the application of tens of volts of bias to remove charges effectively (Chen et al., 2017). In place to evade extremely voltage spikes as well as largely owing, light trapping techniques provide an intriguing option for enhancing CTS absorption significantly. Lately, the utilization of metal holes has resulted in a remarkable photoresponse in OPDs that exceeds the band structure of the constituent materials ((Koch et al., 1982). Utilizing optical micro-cavities, the detecting wavelength may be modified by precisely regulating the cavity size or the depth of the active region. Whereas these gadgets display great results, the frequency response is highly dependent on layer thickness; hence, they may not have been suitable for processes like roll-to-roll, wherein thickness typically varies within a tolerable range (Koch et al., 1982; Zavelani et al., 2006).

In parallel to single-molecule absorbance, NIR detection may also use charge carrier exciton absorption. In 2008, Yang et al. investigated the NIR charge transfer excited-state absorption as well as photocurrent responsiveness of a way to solve bulk heterojunction predicated on P3HT: PCBM. Underneath a high reverse bias, the charge carrier excitons produced between the donor as well as acceptor increased the absorbance range of wavelength to 950 nm, as well as the EQE was 60% at 750 nm. Without even a low-band-gap substance, the great sensitivity of NIR OPDs encompassing 650 to 1,000 nm was achieved. Recent research by Siegmund et al. demonstrates narrowband recognition with spectral widths as small as 36 nm as well as resonance wavelengths between 810 and 1,550 nm, as seen in Figure 6.9.

It is important to note that the infusion of electron-hole pairs from wires into the active layer will result in a substantial fault current underneath a strong reverse bias, which would be exacerbated for organic NIR-absorption substances with a small bandwidth. Such leakage current diminishes the sensitivity. To prevent this leakage current, NIR OPDs often use an injection physical barrier (Chen et al., 2017; Zhang et al., 2019). However, the injection outer layer diminishes the EQE of gadgets as well. To achieve NIR OPDs with high sensitivity, it is required to identify an injection protective barrier that can reconcile the conflict among dark current and EQE.

6.4.4. Panchromatic Organic Photodetectors (OPDs)

OPDs having a sensitivity from UV-Vis to NIR have lately attracted substantial interest for their potential industrial and academic uses. OPDs may also operate in visible light.

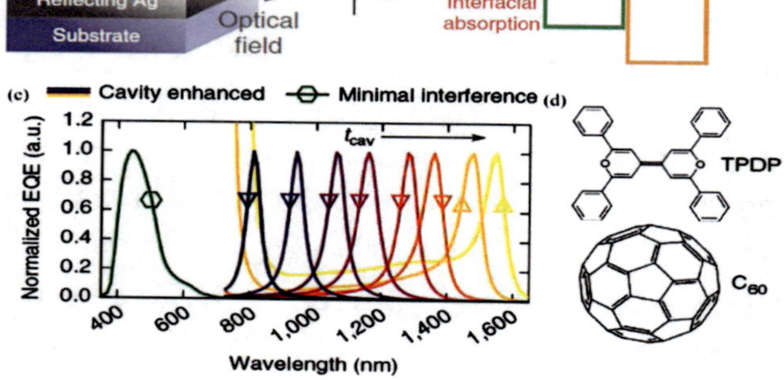

Figure 6.9. (a) Diagram of the device design of a cavity-enhanced organic photodetector, together with an illustration of the optical near field at the resonant

frequencies in the NIR. (b) Energy graph modified for an open circuit. A photon with much less frequency than the optic gap and at minimum the CT state energy (ECT) that is captured at the interface of an electron-donating semiconductor and C_{60} as the receiver. (c) Normalized equivalent quantum efficiency (EQE) spectra of multiple tetraphenyl dipyranylidene:fullerene (TPDP:C_{60}) photodetectors in short circuit. The minimally interfering sample exhibits little absorbance for wavelengths longer than 700 nm. By changing the concentration of the information transmitted and the TPDP: C_{60} mix, the resonant frequency of cavity-enhanced gadgets may be altered from 810 to 1,550 nm. (d) Structures molecules de TPDP et C_{60}.

Source: https://www.nature.com/articles/ncomms15421.

OPDs can not only work in ultraviolet visible, and near-infrared ranges but also achieve a wide spectral range response. In 2009, Forrest and coworkers showed an organic-inorganic hybrid photodetector sensitive to a broad range. The hybrid photodetector employed a carbon nanotube hybrid substance made from particles as well as polymers, resulting in a gadget with a reaction times of just 7.2 ns and a particular detectivity of higher than 1,010 Jones in the range of wavelength of 400 to 1,450 nm. Later, Gong et al. published an all-organic polymer photodetector with a broad spectral range. As seen in Figure 6.10, panchromatic OPDs were produced utilizing a tiny band-gap polymer, poly (5,7-bis(4-decanal-2-thienyl)-thieno(3,4-b) diathiazole-thio-phene-2,5), mixed with PCBM, and demonstrated a wide response spectrum from 300 to 1,450 nm with a particular detectivity more than 1,012 Jones. In 2012, Matthew Menke et al. created layered OPDs by combining three giver substances with comparable light absorptions (Kakavelakis et al., 2016), so overcoming the constraint of the absorbance variety of high levels in single-layer devices as well as attaining a UV-to-NIR spectral reply from 300 to 1,100 nm. Utilizing a layered structure aids in adjusting the spectral response spectrum for each subband. Han et al. announced recently broadband polymer photodetectors with photoresponsivity in the range of 300–1,600 nm. Polymer photodetectors incorporating donor-acceptor polymers with both the donor component dithienobenzotrithiophen (DTBTT) and a thienoisoindigo-based acceptor displayed a specific detectivity larger than 1,011 Jones in the 300–1,300 nm spectra and greater than 1,010 Jones in the 1,300–1,600 nm spectra (Gu et al., 2011; Kakavelakis et al., 2016).

In Table 6.1, we outline the development of OPDs based on their spectral regions. Table 6.1 also includes retail silicon photodetectors as well as

InGaAs photodetectors from Hamamatsu to assist users to comprehending the present level of OPDs.

6.5. A GAIN IN ORGANIC PHOTODETECTORS (OPDS)

Whenever a photon strikes a silicon photodiode, it could only make one electron-hole pair. Avalanche photodiodes (APDs), on the other hand, have internal gaining systems that allow the value of current to be substantially amplified inside the gadget, making the signal more sensitive. The built-in gain reduces the need for separate high-precision amplifier circuits and therefore is appropriate for detecting weak photons. When there is a large reverse-biased voltage is applied to the PN junction of APDs, a very high electric field is generated in the junction, and the photogenerated carriers obtain sufficient energy from the high electric field to collide with the crystal lattice atoms and generate new electron–hole pairs

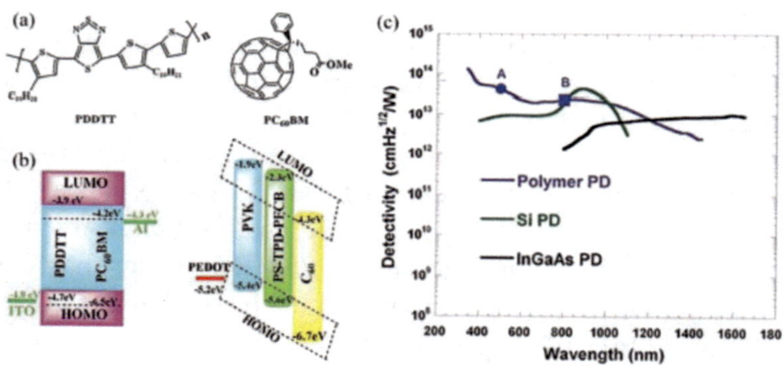

Figure 6.10. *PDDTT and PC60BM molecules. (a) PDDTT, PC60BM, PVK, PS-TPD-PFCB, C_{60}, ITO, PEDOT, and Al energy-level schematics, as well as the structure of the gadget. (b) Particular detectives vs. wavelength of a Si photodetector, InGaAs photodetector, and polymer photodetector.*

Source: https://www.researchgate.net/figure/Device-architectures-and-quantum-efficiency-spectra-for-SubPc-C60-planar-OPVs-a_fig4_331659390.

When a strong electric field is supplied to an APD's PN junction, a very large voltage field is formed, and the photogenerated particles get enough energy from the high electromagnetic current to collision with crystal lattice atoms and create extra two electrons. By subsequent atomic collision, the freshly formed electron-hole pairs might create other electron-hole pairs (Figure 6.11), and the procedure continues to reach photogenerated carrier avalanche multiplication (Van Le et al., 2018; Irandoost and Soleimani-Amiri, 2020).

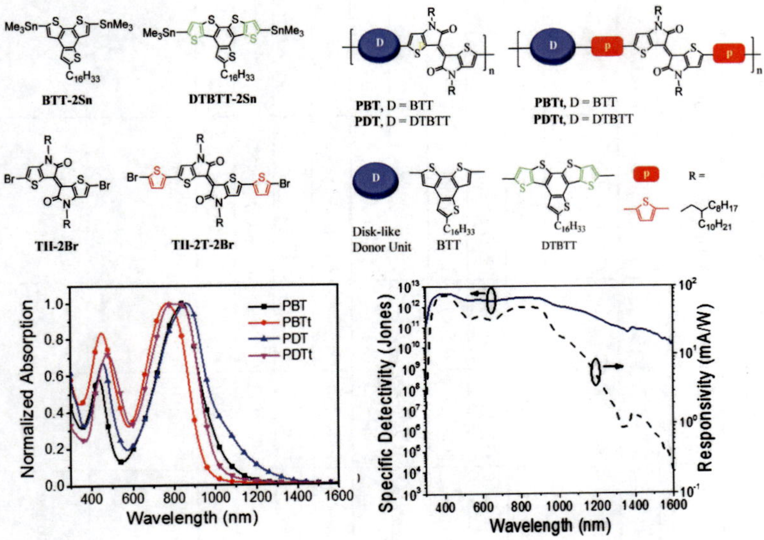

Figure 6.11. *(*a) Polymers produced from disk-like BTT/DTBTT donors and thienoisoindigo (TII) acceptor monomers. (b) Polymer absorption spectra as films, normalized. (c) Measured the device's particular detectivity (D*) and responsivity under 0.1 V using PDT and PC61BM.

Source: https://www.researchgate.net/figure/Donor-acceptor-polymers-derived-from-disk-like-BTT-DTBTT-donor-and-TII-acceptor-monomers_fig1_310837197.

Table 6.1. Photodetectors are Made from Organic Semiconductors Plus Marketable Inorganic Compounds

	Working Wave length	Active Material	A[nm]	Bias[V]	Responsivity [mA W^{-1}]	J dark[A cm^{-2}]	D*[Iones]	Year
Organic Material	Ultraviolet	m-MTDATA Gaq_3	275-475	-8	338@365nm	6×10^{-7}	/	2008
		PFH/NSN	200-425	-12	696@365nm	1.4×10^{-6}	/	2011
	Near infrared	Zn-porphyrine-dimet PCBM, and bipy	900-1600	0	32.8@1130nm 119@1400nm	/	8.8×10^{11}@1130nm 8.0×10^{10}@1400nm	2011
		PDPFSDTPS and $PC_{0D}9M$	300-1200	0	-403@1000nm	/	/	2014
	Visible	P3HT:PCBM	350-750	-0.5	320@500nm	-10^{9}	$>1\times10^{13}$	2012
		p-DTS$(FBTTH_2)_2$ and $PC_{71}BM$	300-800	-0.5	380@700nm	1.1×10^{-10}	9.2×10^{12}@700nm	2014
		PCDTBT and $PC_{77}BM$	400-700	-5	302@500nm	7.4×10^{-10}	2.1×10^{13}@532nm	2015
	UV-NIR	PDDTT.$PC_{50}BM$	300-1450	-0.1	>167@500nm	$<10^{-9}$	2.3×10^{13}@800nm	2009
		1-BF and C_{60}	400-1440	0	2.1%(EQE peak)	/	3.7×10^{9}@1000nm	2016
		1-TPFB and C_{60}	400-1460	0	1.1%(EQE peak)	/	5.3×10^{10}@1000nm	
		2-BF4 and C_{60}	400-1590	0	1.4%(EQE peak)	/	7×10^{9}@1000nm	
		PDT.$PC_{51}BM$	300-1600	-0.1	2.4@900nm	1.38×10^{-10}	2.6×10^{12}@900nm	2017
		4.7dibromo-[1.2.5]selenadiazolo [3.4-c]pyridine and $PC_{71}BM$	450-1650	0	-58@1200nm	/	2×10^{11}@1330nm	2017
		3b with 4,6bis[5-bromo-2-thieny] thieno[3,4-c][1.2.5]thiadiazole and $PC_{71}BM$	450-1800	0	-3@1200nm	/	1.2×10^{8}@1800nm	

Inorganic material	Visible-NIR	Si (S1336-18BK)	320-1100	-0.01	500@960nm	$<1.7\times10^{-9}$	2×10^{13}@960nm	2018
	UV-NIR	Si (S1336-18BQ)	190-1100	-0.01	120@200nm 500@960nm	$<1.7\times10^{-9}$	2×10^{13}@960nm	
	Visible	Si(8265)	340-720	-1	300@540nm	$<3\times10^{-10}$	/	
	Visible-NIR	InGaAs(G12180-03K)(noncooled)	500-1700	-1	220@650nm 1000@1550nm	$<2\times10^{-7}$	5×10^{12}@1550nm	2017
	NIR	InGaAs(G12180-030A)	900-1700	-1	1100@1550nm	3.5×10^{-8}	6.3×10^{12}@1550nm	
	NIR	InGaAs(G12183-130K)(TE-cooled)	900-2570	-0.5	1200@2300nm	1.3×10^{-4}	3×10^{11}@2300nm	

Furthermore, given the limited density and high exciton bond length of organic compounds, this mechanism is exceedingly improbable in OPDs. Despite this, photocurrent multiplication was seen in OPDs. Hiramoto et al. demonstrated photocurrent multiplication in OPDs with the configuration of a perylene pigment screen among Au electrodes for the first time in 1994. The photocurrent amplification in the OPD was aided by a system that depends on the tunneling injection of particles from the Au poles to the perylene, which was induced by photo accumulated space charge polarization of entrapped gaps at the organic/metal contact. The scientists further discovered photocurrent multiplication at the heterojunction of p-type phthalocyanine (CuPc) and n-type perylene (Me-PTC) pigment films (Smythe et al., 2007; Bhatia and Gupta, 2011). In 2006, Reynaert et al. demonstrated a photocurrent doubling in the majority of a single-phase disorganized organic semiconductor using an exciton power transmission method to trapped particles to improve photoconductivity in OPDs (Smythe et al., 2007). At the moment, the two methods outlined above are primarily employed in OPDs to produce photocurrent doubling. The multiplication process of exciton transfer of energy to entrapped radicals in bulk counterparts has the added benefit of not requiring exciton breakup at a donor/acceptor contact, which is important for producing NIR OPDs with amplification. Furthermore, light-induced energy injection or transit out of a sustainably grown or natural contact to an organic/organic connection (Neff et al., 2006; Singh et al., 2013).

The presence of field-activated physical trapping at the junction aids in the decrease of dark current density in increasing OPDs. Following research, the performance measures of OPDs with rising have been greatly enhanced, with some exceeding those of typical inorganic photodetectors. Chen et al. disclosed in 2008 a hybrid photodetector based on a nanoparticle-assisted photosensitive gain in polymer and fullerene composites. Around 4.5 V and a wavelength of 350 nm, and EQE of up to 8,000% were obtained (Figure 6.12) (Fujikawa et al., 2008).

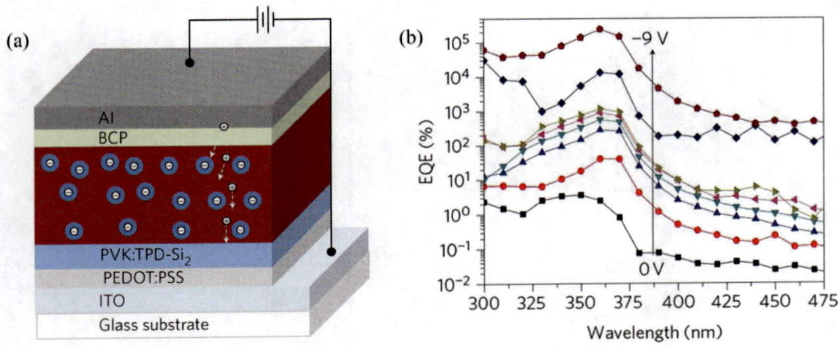

Figure 6.12. (a) Nanocomposite UV photodetector device construction; (b) The UV photodetector's EQE with PVK.

Source: https://www.researchgate.net/figure/a-The-schematic-diagram-of-the-PD-device-b-Noise-characteristics-of-the-PVK-ZnO-PDs_fig7_339379825.

Without the use of D-A exciton dissociating materials, a NIR organic photodetector with gain at a low bias voltage was developed in 2009.

Peak EQE values in devices using PVK: ZnO and P3HT: ZnO active layers were 245 × 300% and 340 × 600%, respectively. With effective regulation of the dark current density, the devices demonstrated a specific detectivity of 3.4 × 1,015 Jones at 360 nm, outperforming conventional inorganic UV photodetectors. Aside from their strong responsivity and detectivity, the OPDs had a good linear response from 1,109 W cm^2 to 1,101 W cm^2. Zhou et al. have created ultrahigh-gain polymer photodetectors with a spectrum response from UV to NIR employing ZnO nanoparticles as an anode interfacial layer. The devices have an extraordinarily low gain onset voltage of 1.5 mV after 30 seconds of UV-light radiation. A very high EQE of 140,000% was also reached with a modest bias of 0.5 V (Figure 6.13 and Table 6.2).

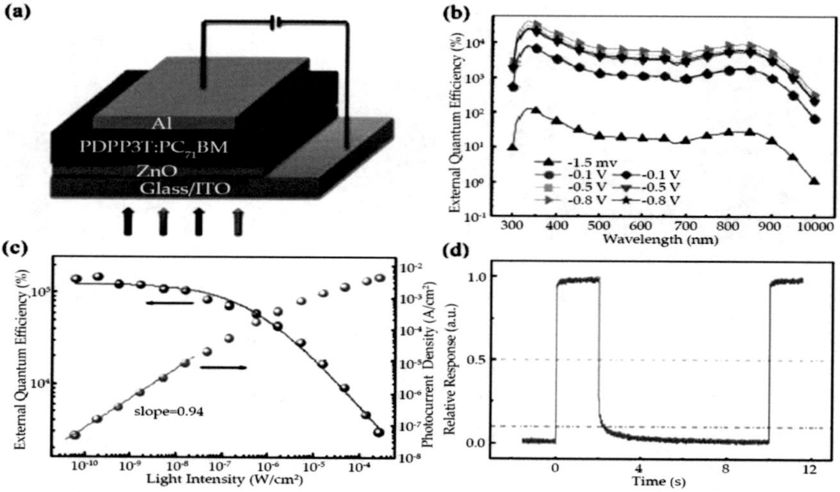

Figure 6.13. (a) The device structure's schematic arrangement. (b) EQE spectra obtained following UV-light treatment at various bias voltages. (c) Incident power intensity-photocurrent density as well as EQE parameters of the resultant photoconductor following UV-light treatment at –0.5 V bias voltage. The dense line is the fitted curve based on the photoconductive prototype. (d) The transient photocurrent normalized at 0.1 Hz.

Source: https://pubs.acs.org/doi/10.1021/acsomega.0c04136.

Table 6.2. Major Advances in Organic Photodetectors with Improvement

Year	EQE	Bias	A Mechanism[') [nm]	Location of traps	Interfacial band bending	1)• pones]	Jerk (mA cm-9	LDR [W	Rise time	Fall time
1994	-10000%	16	350-700 TACCTI	Hole traps in Me-PTC.	Au/Me-PTC	/	50 @16V	/	/	/
	(IQE)									
1998	3000% (IQE)	>20	350-800 TACCTI	Hole traps in CuPc/ Me-PTC	CuPc/Me-PTC	/	/	/	Several ten second[d]	n10 VI) r ,
2000	1.7×10^3	20	/ TACCTI	Trapping sites at	NTCDA/Au	/	0.12 @18V	/	3.7 s	/ C

	(IQE)			NTCDA/Au interface						
2009	-10 (gain)	5	400-1100 PG	OSnNcC12	/	/	-,1 (0-5 VII	/	/	/ [
2010	32 500%	-4	400-800 TACCTI	Hole traps in co near	ITO/NTCDA	/	-,1 0-4 VII	/	/	/ [
				NTCDA						
2012	340 600%	-9	300-700 TACCTI	Electron traps in nanoparticles	AI/BCP	3.4 x $10^{19:9}$	..10^4 tb-9 V')	10'- 10^{-9}	25 ps	558 its [
2012	-.550096	-3.7	400-1200 PG	Electron traps at the	/	/	-'1 0-3 VII	/	/	/ C
				NIR dopants						
2014	408%	-8	300-700 TACCTI	Hole traps in C_{93} and nanoparticles	C-TPD1n0	6.5 x 10^{176})	n10^{-2} (d)-8 V	10'- 10^{13}	/	/ C
2015	37 500%	-25	-330-830 TACCTI	Electrons trapped in	PTB7-Th/AI	1.91 x lei	/	/	-1 5^d)	n1 s^dI [
				$PC_{,1}BM$						
2015	112100%	-25	350-750 TACCTI	Electrons trapped in	P3HT/AI	5.0 x 10"4	0.16 @ -25V	/	41 sd)	/ C
				$PC_{,1}BM$						
2016	-10000%	-10	250-1000 TACCTI	Hole traps in C_w	ITO/TPBi	4 x 10"4	1 0-10V	/	/	/ [
2016	140000%	-0.5	300-1000 PG	/	/	6.3 x 104	46 ig-0.5 V	/	2.8 ms	0.27 s[
2017	1.59 x $10^3$96	1	-350-900 TACCTI	/	PBDTT-DPP/	',le')	/	10- $10^{1°}$	1621is	7.9 ms[
	(10 V)				Mo03					

The photocurrent improvement process was attributed to the rapid transit and replacement of photogenerated electrons during their extended lifespan, and the new findings were consistent with the suggested photoconductive design. Over the last two eras, a great deal of effort has gone towards increasing the EQE, lowering the dark current density and operating voltage, broadening the spectrum reply, as well as improving the reply speed of OPDs with gain (Fen and Yunus, 2011). Table 6.1 summarizes a few of the important benchmarks.

6.6. LINEAR DYNAMIC RANGE (LDR)

The extent whereby the output voltage retains a linear connection with the original signal while a photodetector is lit is described by the linear dynamic range (LDR), which is another key quality characteristic of photodetectors. In other words, a photodetector's output signal is proportional to the quantity of light it receives inside the selected timeframe. A photodetector's photo responsiveness is constant throughout the area known as the linear region. The reaction detracts from linearity and becomes less responsive as the incoming light intensity rises. The kind of gadget, as well as the operational circuit, affect the linearity's range. The NEP, which is associated with the OPD's work mode, establishes the bottom limit of an OPD's linearity. The lower threshold drops as a result of the decreased gadget distortion in the photovoltaic state compared to the photoconductive state. The highest limit is determined by the load resistor, series resistance, as well as decrease in excitons' use after absorption. Whenever the series resistance element rises, the linearity declines.

Since the resistance is unavoidable, it is necessary to minimize electrode resistance to improve LDR. A bright light environment reduces an OPD's quantum efficiency because it is less effective in collecting free carriers. For instance, the space-charge-limited photon energy, that is frequently observed in organic solar cells and OPDs, would hinder the performance of the device if the electron/hole circulations are not balanced. Resulting in greater charge collecting effectiveness for OPDs, the supplied resistor causes a slight rise in photocurrent as well as a rise in the maximum bound. (Mahmudin et al., 2015) As of right now, the finest LDR in OPDs can span more than 10 orders of magnitude that is equivalent to inorganic Si photodetectors available for purchase (Mahmudin et al., 2015).

6.7. RESPONSE SPEED

The reaction speed is a crucial characteristic for photodetectors, particularly in optical communication system uses, since it specifies how quickly the photodetectors react to events in the incoming optical signal. The response speed is described in the time-frequency domain by the reaction times. The speed of reaction is a variable that indicates how quickly a photodetector reacts to a change in the optical original indication (Figure 6.14).

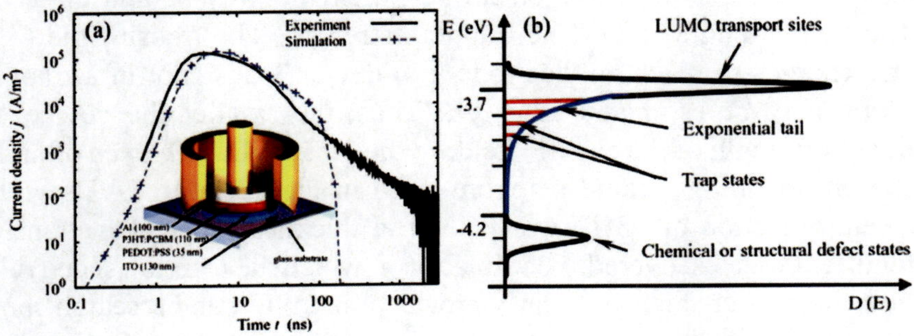

Figure 6.14. (a) Modeling of the transient photoresponse lacking trap conditions vs. experiment results of a P3HT: PCBM photodetector. The inset depicts the experimental device's conceptual construction. (b) Diagrammatic representation of electron transport pathways in a bi-modal state pattern. Trap-free transit locations are represented by the narrow Gaussian distribution focused on the valance band. Inside the band gap, the tail points function as a constant exponentially trap distribution. A single deep trapping layer also symbolizes chemical contaminants or structural faults.

Source: https://onlinelibrary.wiley.com/doi/abs/10.1002/aenm.201601574.

The factor includes the proportional gain and falling edge, in which the increased time is described as the break wherein the output electrical impulse shifts from 10% to 90% of the stable highest benefit, and the fall time is originally scheduled wanted for the production of electric signals to alter from 90% to 10% of the stable highest benefit. The bandwidth (or cutoff frequency) characterizes the dynamic response in the frequency response, that is a crucial factor for determining the temporal reaction of a photodetector to a sinusoidally modified transmitted fiber. Particularly, it is described as the frequency band across which a signal's frequency responsiveness declines by 3 dB relative to its stable optimum amount.

In OPDs, exciton diffusion to the donor/acceptor contacts and charged collection take far longer than exciton creation from light absorption and charge transport exciton dissociation. Like inorganic photodetectors, the reaction time of an OPD is influenced by the time required for exciton propagation and charging storage. Furthermore, the distinctive RC time constant influences the OPDs' reaction speed. Owing to the intricacy of organic compounds and gadget architectures in OPDs, though, the parameters that influence reaction speed are difficult.

In OPDs, the collecting duration of photogenerated carriers is determined by their distance (d) from the electrodes, their mobility (), and either the electric field strength (E) or the trapping strength. The transit time () for electric charges to reach an electrode in a device that is not in a trapping condition may be estimated as = d/E. Thus, the smaller the device, the greater the strength of the electric field, so the greater the influence of highly mobile on obtaining a rapid reaction speed in the OPD. In OPDs with a bulk heterojunction of P3HT and PCBM, a thickness-dependent transient photocurrent was discovered. The situation in which the reaction speed relies on the voltage source is particularly prevalent in OPDs, and research shows that the bandwidth displays a roughly rising trend as the supplied voltage is increased. The electric field intensity, temperature, as well as density of the electric charge would affect the movement of the electric charge. Consequently, the temperature, as well as brightness of incoming light, as well as brightness of incoming light should also influence the reaction speed of OPDs. The effect of temperature on the instantaneous photocurrent concentration features of organic solar cells as well as OPDs was examined by Zufle et al. The researchers discovered that the rate of decrease in current densities after reaching the ultimate value accelerates from 11 to 50 C and that the electronic structure almost doubles with rising temperature, making it more appropriate for the analysis of high transmissions. In OPDs, the time required for exciton propagation to the donor/acceptor interface is typically proportional to the exciton diffusion coefficient (D), lifespan (τ), and device structure. Given that the average lifespan of Frenkel excitons in semiconductor materials is on the scale of nanoseconds, the exciton diffusion length (Ld) is usually brief (Srivastava and Gupta, 2011).

Morimune et al. observed the phenomena of carrier frequency restricted by the ion diffusion distance at a higher frequency in a multilayer OPD composed of CuPc and BPPC. For an OPD exciting triplet excitation energy, the response speed is significantly constrained by the dispersion of excitons. Since a bulk heterojunction minimizes the length of exciton diffusion, this design is often used to increase reaction speed. Due to the decreased mobility of heterojunctions in bulk, acquire a rapid reaction.

In addition to the above-mentioned parameters, traps may also have a significant impact on the twitch reflexes of OPDs. Also reported to happen in OSC are dispersive start charging transportation and a hyperbolic trap. By contrasting research observations along with or not backdrop lighting in polymer solar cells, McNeill et al. discovered indication for the presence of deep trick conditions as the origin of the photoresponse on a microsecond timeframe. Christ et al. conducted an arithmetical recreation of multiple-trapping based on drift-diffusion to clarify the long tail of the transient photoresponse in organic solar cells and photodetectors. The findings demonstrated that the traditional drift-diffusion calculation with an actual movement for every charge carrier is insufficient to simulate the photoresponse of organic gadgets, and that trap positions must be taken into account. In addition, Arca et al. investigated the effect of traps at the interfaces of the spray-coated bulk heterojunction of P3HT: PCBM and the electrodes. The scientists evaluated the active performance of OPDs at pulsed light concentrations and discovered that traps at the stack interfaces, as opposed to those in the majority of the dynamic layer, had a significant effect on the reaction speed of OPDs.

Because of the thin charge transport groups and the fact that charge transport is driven by heated triggered hopping among extremely localized sites (or traps), organic optoelectronic materials have lower mobility than inorganic semiconductors. Though, the optical absorption coefficient of organic resources is often much greater than that of inorganic materials, allowing the dynamic coating of OPDs to be as thin as a few 100 nanometers. So, the bandwidth of OPDs may vary from a few kilohertz to several 100 megahertz, which is described alongside the ranges of silicon PIN photodetectors from Hamamatsu (Barakat et al., 2008).

6.8. CONCLUSION

Although OSC offer several benefits over inorganic semiconductors, OPDs have not been deployed in practice. A single probable explanation is that the general presentation of OPDs is up till now lesser than that of inorganic photodetectors, and the benefits of OSC, like elasticity, big area, as well as cheap cost, will not be completely realized in the upcoming days. Though, there is still space for change in OPDs' basic research performance. Exploration of the key functioning mechanism and changes to performance might assist hasten the industrialization process.

REFERENCES

1. Barakat, N. A., Woo, K. D., Kanjwal, M. A., Choi, K. E., Khil, M. S., & Kim, H. Y., (2008). Surface plasmon resonances, optical properties, and electrical conductivity thermal hystersis of silver nanofibers produced by the electrospinning technique. *Langmuir, 24*(20), 11982–11987.
2. Bässler, H., & Köhler, A., (2011). Charge transport in organic semiconductors. *Unimolecular and Supramolecular Electronics I*, 1–65.
3. Bhatia, P., & Gupta, B. D., (2011). Surface-plasmon-resonance-based fiber-optic refractive index sensor: Sensitivity enhancement. *Applied Optics, 50*(14), 2032–2036.
4. Brédas, J. L., & Marder, S. R., (2016). *The WSPC Reference on Organic Electronics: Organic Semiconductors* (Vol. 1). Singapore: World Scientific.
5. Brédas, J. L., Calbert, J. P., Da Silva, F. D. A., & Cornil, J., (2002). Organic semiconductors: A theoretical characterization of the basic parameters governing charge transport. *Proceedings of the National Academy of Sciences, 99*(9), 5804–5809.
6. Bronstein, H., Nielsen, C. B., Schroeder, B. C., & McCulloch, I., (2020). The role of chemical design in the performance of organic semiconductors. *Nature Reviews Chemistry, 4*(2), 66–77.
7. Brütting, W., (2005). Introduction to the physics of organic semiconductors. *Physics of Organic Semiconductors*, 1–14.
8. Cai, Y., Qin, A., & Tang, B. Z., (2017). Siloles in optoelectronic devices. *Journal of Materials Chemistry C, 5*(30), 7375–7389.
9. Chang, C. R., & Jean, J. H., (1999). Crystallization kinetics and mechanism of low-dielectric, low-temperature, cofirable $CaO-B_2O_3-SiO_2$ glass-ceramics. *Journal of the American Ceramic Society, 82*(7), 1725–1732.
10. Chang, E., Bussi, G., Ruini, A., & Molinari, E., (2004). Excitons in carbon nanotubes: An ab initio symmetry-based approach. *Physical Review Letters, 92*(19), 196401.
11. Chen, J., Perebeinos, V., Freitag, M., Tsang, J., Fu, Q., Liu, J., & Avouris, P., (2005). Bright infrared emission from electrically induced excitons in carbon nanotubes. *Science, 310*(5751), 1171–1174.
12. Chen, P., Xiong, Z., Wu, X., Shao, M., Meng, Y., Xiong, Z. H., & Gao, C., (2017). Nearly 100% efficiency enhancement of

$CH_3NH_3PbBr_3$ perovskite light-emitting diodes by utilizing plasmonic Au nanoparticles. *The Journal of Physical Chemistry Letters, 8*(17), 3961–3969.

13. Chen, Y., Fang, L., Yi, D., Li, X., & Hong, X., (2020). Thermo-optic property measurement using surface plasmon resonance-based fiber optic sensor. *IEEE Sensors Journal, 20*(19), 11357–11363.
14. Damos, F. S., Luz, R. C., & Kubota, L. T., (2006). Investigations of ultrathin polypyrrole films: Formation and effects of doping/dedoping processes on its optical properties by electrochemical surface plasmon resonance (ESPR). *Electrochimica Acta, 51*(7), 1304–1312.
15. Denton, G. J., Tessler, N., Harrison, N. T., & Friend, R. H., (1997). Factors influencing stimulated emission from poly (p-phenylenevinylene). *Physical Review Letters, 78*(4), 733.
16. Dong, H., Wang, C., & Hu, W., (2010). High performance organic semiconductors for field-effect transistors. *Chemical communications, 46*(29), 5211–5222.
17. Eastham, J., (2000). Designing out failure in composites: One day seminar. *Materials Technology, 15*(1), 96–98.
18. Fen, Y. W., & Yunus, W. M. M., (2011). Characterization of the optical properties of heavy metal ions using surface plasmon resonance technique. *Opt. Photonics J., 1*(03), 116–123.
19. Freitag, M., Chen, J., Tersoff, J., Tsang, J. C., Fu, Q., Liu, J., & Avouris, P., (2004). Mobile ambipolar domain in carbon-nanotube infrared emitters. *Physical Review Letters, 93*(7), 076803.
20. Freitag, M., Martin, Y., Misewich, J. A., Martel, R., & Avouris, P., (2003). Photoconductivity of single carbon nanotubes. *Nano Letters, 3*(8), 1067–1071.
21. Fujikawa, R., Baryshev, A. V., Kim, J., Uchida, H., & Inoue, M., (2008). Contribution of the surface plasmon resonance to optical and magneto-optical properties of a Bi: YIG-Au nanostructure. *Journal of Applied Physics, 103*(7), 07D301.
22. Gu, X., Qiu, T., Zhang, W., & Chu, P. K., (2011). Light-emitting diodes enhanced by localized surface plasmon resonance. *Nanoscale Research Letters, 6*(1), 1–12.
23. Guldi, D. M., Taieb, H., Rahman, G. A., Tagmatarchis, N., & Prato, M., (2005). Novel photoactive single-walled carbon nanotube–porphyrin

polymer wraps: Efficient and long-lived intracomplex charge separation. *Advanced Materials, 17*(7), 871–875.

24. Guo, X., Huang, L., O'Brien, S., Kim, P., & Nuckolls, C., (2005). Directing and sensing changes in molecular conformation on individual carbon nanotube field effect transistors. *Journal of the American Chemical Society, 127*(43), 15045–15047.
25. Hayes, G. R., Samuel, I. D. W., & Phillips, R. T., (1995). Exciton dynamics in electroluminescent polymers studied by femtosecond time-resolved photoluminescence spectroscopy. *Physical Review B, 52*(16), R11569.
26. Hecht, D. S., Ramirez, R. J., Briman, M., Artukovic, E., Chichak, K. S., Stoddart, J. F., & Grüner, G., (2006). Bioinspired detection of light using a porphyrin-sensitized single-wall nanotube field effect transistor. *Nano Letters, 6*(9), 2031–2036.
27. Heliotis, G., Xia, R., Bradley, D. D. C., Turnbull, G. A., Samuel, I. D. W., Andrew, P., & Barnes, W. L., (2003). Blue, surface-emitting, distributed feedback polyfluorene lasers. *Applied Physics Letters, 83*(11), 2118–2120.
28. Henson, Z. B., Müllen, K., & Bazan, G. C., (2012). Design strategies for organic semiconductors beyond the molecular formula. *Nature Chemistry, 4*(9), 699–704.
29. Herrick, R. W., & Guo, Q., (2021). Introduction to optoelectronic devices. In: *Reliability of Semiconductor Lasers and Optoelectronic Devices* (pp. 1–45). Woodhead Publishing.
30. Hlubina, P., Kadulova, M., Ciprian, D., & Sobota, J., (2014). Reflection-based fiber-optic refractive index sensor using surface plasmon resonance. *Journal of the European Optical Society-Rapid Publications, 9.*
31. Hofmann, A. I., Kroon, R., & Müller, C., (2019). Doping and processing of organic semiconductors for plastic thermoelectrics. In: *Handbook of Organic Materials for Electronic and Photonic Devices* (pp. 429–449). Woodhead Publishing.
32. Irandoost, R., & Soleimani-Amiri, S., (2020). Design and analysis of high efficiency perovskite solar cell with ZnO nanorods and plasmonic nanoparticles. *Optik, 202*, 163598.
33. Islam, M. F., Milkie, D. E., Kane, C. L., Yodh, A. G., & Kikkawa, J. M., (2004). Direct measurement of the polarized optical absorption

cross section of single-wall carbon nanotubes. *Physical Review Letters, 93*(3), 037404.

34. Itkis, M. E., Borondics, F., Yu, A., & Haddon, R. C., (2006). Bolometric infrared photoresponse of suspended single-walled carbon nanotube films. *Science, 312*(5772), 413–416.
35. Jacob, M. V., (2014). Organic semiconductors: Past, present and future. *Electronics, 3*(4), 594–597.
36. Jiang, H., & Kloc, C., (2013). Single-crystal growth of organic semiconductors. *MRS Bulletin, 38*(1), 28–33.
37. Jiang, W., Li, Y., & Wang, Z., (2013). Heteroarenes as high performance organic semiconductors. *Chemical Society Reviews, 42*(14), 6113–6127.
38. Junaid, M., Md Khir, M. H., Witjaksono, G., Ullah, Z., Tansu, N., Saheed, M. S. M., & Siddiqui, M. A., (2020). A review on graphene-based light emitting functional devices. *Molecules, 25*(18), 4217.
39. Kakavelakis, G., Vangelidis, I., Heuer-Jungemann, A., Kanaras, A. G., Lidorikis, E., Stratakis, E., & Kymakis, E., (2016). Plasmonic Backscattering Effect in High-Efficient Organic Photovoltaic Devices. *Advanced Energy Materials, 6*(2), 1501640.
40. Karplus, M., & Porter, R. N., (1970). Atoms and molecules; an introduction for students of physical chemistry. *Atoms and Molecules; An Introduction for Students of Physical Chemistry*, 1–20.
41. Knupfer, M., (2003). Exciton binding energies in organic semiconductors. *Applied Physics A, 77*(5), 623–626.
42. Koch, T. L., Chiu, L. C., & Yariv, A., (1982). Analysis and performance of a picosecond dye laser amplifier chain. *Journal of Applied Physics, 53*(9), 6047–6059.
43. Köhler, A., & Bässler, H., (2009). Triplet states in organic semiconductors. *Materials Science and Engineering: R: Reports, 66*(4–6), 71–109.
44. Krause, S., Casu, M. B., Schöll, A., & Umbach, E., (2008). Determination of transport levels of organic semiconductors by UPS and IPS. *New Journal of Physics, 10*(8), 085001.
45. Larrimore, L., Nad, S., Zhou, X., Abruna, H., & McEuen, P. L., (2006). Probing electrostatic potentials in solution with carbon nanotube transistors. *Nano Letters, 6*(7), 1329–1333.

46. Lee, J. U., (2005). Photovoltaic effect in ideal carbon nanotube diodes. *Applied Physics Letters, 87*(7), 073101.
47. Lee, J. U., (2007). Band-gap renormalization in carbon nanotubes: Origin of the ideal diode behavior in carbon nanotube p– n structures. *Physical Review B, 75*(7), 075409.
48. Leonard, F., & Stewart, D. A., (2006). Properties of short channel ballistic carbon nanotube transistors with ohmic contacts. *Nanotechnology, 17*(18), 4699.
49. Léonard, F., & Tersoff, J., (2002). Dielectric response of semiconducting carbon nanotubes. *Applied Physics Letters, 81*(25), 4835–4837.
50. Li, R., Hu, W., Liu, Y., & Zhu, D., (2010). Micro-and nanocrystals of organic semiconductors. *Accounts of Chemical Research, 43*(4), 529–540.
51. Liang, Z., Sun, J., Jiang, Y., Jiang, L., & Chen, X., (2014). Plasmonic enhanced optoelectronic devices. *Plasmonics, 9*(4), 859–866.
52. Lin, S. Y., Tsai, Y. R., & Lee, S. C., (2001). High-performance InAs/GaAs quantum-dot infrared photodetectors with a single-sided Al 0.3 Ga 0.7 As blocking layer. *Applied Physics Letters, 78*(18), 2784–2786.
53. Lin, Y., Fan, H., Li, Y., & Zhan, X., (2012). Thiazole-based organic semiconductors for organic electronics. *Advanced Materials, 24*(23), 3087–3106.
54. Liu, F., & Nunzi, J. M., (2012). Enhanced organic light emitting diode and solar cell performances using silver nano-clusters. *Organic Electronics, 13*(9), 1623–1632.
55. Liu, X., Jeong, K. S., Williams, B. P., Vakhshouri, K., Guo, C., Han, K., & Asbury, J. B., (2013). Tuning the dielectric properties of organic semiconductors via salt doping. *The Journal of Physical Chemistry B, 117*(49), 15866–15874.
56. Loucif-Saibi, R., Nakatani, K., Delaire, J. A., Dumont, M., & Sekkat, Z., (1993). Photoisomerization and second harmonic generation in disperse red one-doped and-functionalized poly (methyl methacrylate) films. *Chemistry of Materials, 5*(2), 229–236.
57. Ma, Z., Geng, H., Wang, D., & Shuai, Z., (2016). Influence of alkyl side-chain length on the carrier mobility in organic semiconductors: Herringbone vs. pi–pi stacking. *Journal of Materials Chemistry C, 4*(20), 4546–4555.

58. Mahmudin, L., Suharyadi, E., Utomo, A. B. S., & Abraha, K., (2015). Optical properties of silver nanoparticles for surface plasmon resonance (SPR)-based biosensor applications. *Journal of Modern Physics, 6*(08), 1071.
59. Mann, D., Kato, Y. K., Kinkhabwala, A., Pop, E., Cao, J., Wang, X., & Dai, H., (2007). Electrically driven thermal light emission from individual single-walled carbon nanotubes. *Nature Nanotechnology, 2*(1), 33–38.
60. Marcus, M. S., Simmons, J. M., Castellini, O. M., Hamers, R. J., & Eriksson, M. A., (2006). Photogating carbon nanotube transistors. *Journal of Applied Physics, 100*(8), 084306.
61. Martín, J., Dyson, M., Reid, O. G., Li, R., Nogales, A., Smilgies, D. M., & Stingelin, N., (2018). On the effect of confinement on the structure and properties of small-molecular organic semiconductors. *Advanced Electronic Materials, 4*(1), 1700308.
62. Marty, L., Adam, E., Albert, L., Doyon, R., Menard, D., & Martel, R., (2006). Exciton formation and annihilation during 1D impact excitation of carbon nanotubes. *Physical Review Letters, 96*(13), 136803.
63. Meng, Q., Gao, J., Li, R., Jiang, L., Wang, C., Zhao, H., & Hu, W., (2009). New type of organic semiconductors for field-effect transistors with carbon-carbon triple bonds. *Journal of Materials Chemistry, 19*(10), 1477–1482.
64. Misewich, J. A., Martel, R., Avouris, P., Tsang, J. C., Heinze, S., & Tersoff, J., (2003). Electrically induced optical emission from a carbon nanotube FET. *Science, 300*(5620), 783–786.
65. Myers, J. D., & Xue, J., (2012). Organic semiconductors and their applications in photovoltaic devices. *Polymer Reviews, 52*(1), 1–37.
66. Nakano, M., Osaka, I., Hashizume, D., & Takimiya, K., (2015). α-modified naphthodithiophene diimides- molecular design strategy for air-stable n-channel organic semiconductors. *Chemistry of Materials, 27*(18), 6418–6425.
67. Neff, H., Zong, W., Lima, A. M. N., Borre, M., & Holzhüter, G., (2006). Optical properties and instrumental performance of thin gold films near the surface plasmon resonance. *Thin Solid Films, 496*(2), 688–697.

68. Newman, C. R., Frisbie, C. D., Da Silva, F. D. A., Brédas, J. L., Ewbank, P. C., & Mann, K. R., (2004). Introduction to organic thin film transistors and design of n-channel organic semiconductors. *Chemistry of Materials, 16*(23), 4436–4451.

69. Nie, J., Lindén, L. Å., Rabek, J. F., Fouassier, J. P., Morlet-Savary, F., Scigalski, F., & Andrzejewska, E., (1998). A reappraisal of the photopolymerization kinetics of triethyleneglycol dimethacrylate initiated by camphorquinone-N, N-dimethyl-p-toluidine for dental purposes. *Acta polymerica, 49*(4), 145–161.

70. Nitti, A., Forti, G., Bianchi, G., Botta, C., Tinti, F., Gazzano, M., & Pasini, D., (2021). Anthradithiophene-based organic semiconductors through regiodirected double annulations. *Journal of Materials Chemistry C, 9*(29), 9302–9308.

71. Okamoto, T., Suzuki, T., Tanaka, H., Hashizume, D., & Matsuo, Y., (2012). Tetracene dicarboxylic imide and its disulfide: Synthesis of ambipolar organic semiconductors for organic photovoltaic cells. *Chemistry–An Asian Journal, 7*(1), 105–111.

72. Pecile, C., Palnelli, A., & Girlando, A., (1989). Studies of organic semiconductors for 40 years—V. *Molecular Crystals and Liquid Crystals, 171*(1), 69–87.

73. Perebeinos, V., & Avouris, P., (2007). Exciton ionization, Franz–Keldysh, and stark effects in carbon nanotubes. *Nano Letters, 7*(3), 609–613.

74. Perebeinos, V., Tersoff, J., & Avouris, P., (2004). Scaling of excitons in carbon nanotubes. *Physical Review Letters, 92*(25), 257402.

75. Phillips, J. M., Burrows, P. E., Davis, R. F., Simmons, J. A., Malliaras, G. G., So, F., & Vittitow, M. P., (2006). *Basic Research Needs for Solid-State Lighting. Report of the Basic Energy Sciences Workshop on Solid-State Lighting*. DOESC (USDOE Office of Science (SC)).

76. Pop, E., Mann, D., Cao, J., Wang, Q., Goodson, K., & Dai, H., (2005). Negative differential conductance and hot phonons in suspended nanotube molecular wires. *Physical Review Letters, 95*(15), 155505.

77. Pope, M., & Swenberg, C. E., (1999). *Electronic Processes in Organic Crystals and Polymers* (Vol. 56). Oxford University Press on Demand.

78. Root, S. E., Savagatrup, S., Pais, C. J., Arya, G., & Lipomi, D. J., (2016). Predicting the mechanical properties of organic semiconductors using coarse-grained molecular dynamics simulations. *Macromolecules,*

49(7), 2886–2894.

79. Ruini, A., Caldas, M. J., Bussi, G., & Molinari, E., (2002). Solid state effects on exciton states and optical properties of PPV. *Physical Review Letters, 88*(20), 206403.
80. Saito, R., Dresselhaus, G., & Dresselhaus, M. S., (2000). Trigonal warping effect of carbon nanotubes. *Physical Review B, 61*(4), 2981.
81. Schwartz, B. J., Hide, F., Andersson, M. R., & Heeger, A. J., (1997). Ultrafast studies of stimulated emission and gain in solid films of conjugated polymers. *Chemical Physics Letters, 265*(3–5), 327–333.
82. Schwarze, M., Tress, W., Beyer, B., Gao, F., Scholz, R., Poelking, C., & Leo, K., (2016). Band structure engineering in organic semiconductors. *Science, 352*(6292), 1446–1449.
83. Shen, Y., Hosseini, A. R., Wong, M. H., & Malliaras, G. G., (2004). How to make ohmic contacts to organic semiconductors. *ChemPhysChem, 5*(1), 16–25.
84. Simmons, J. M., In, I., Campbell, V. E., Mark, T. J., Léonard, F., Gopalan, P., & Eriksson, M. A., (2007). Optically modulated conduction in chromophore-functionalized single-wall carbon nanotubes. *Physical Review Letters, 98*(8), 086802.
85. Singh, S., Mishra, S. K., & Gupta, B. D., (2013). Sensitivity enhancement of a surface plasmon resonance based fiber optic refractive index sensor utilizing an additional layer of oxides. *Sensors and Actuators A: Physical, 193*, 136–140.
86. Smythe, E. J., Cubukcu, E., & Capasso, F., (2007). Optical properties of surface plasmon resonances of coupled metallic nanorods. *Optics Express, 15*(12), 7439–7447.
87. Spataru, C. D., Ismail-Beigi, S., Benedict, L. X., & Louie, S. G., (2004). Excitonic effects and optical spectra of single-walled carbon nanotubes. *Physical Review Letters, 92*(7), 077402.
88. Srivastava, S. K., & Gupta, B. D., (2011). Influence of ions on the surface plasmon resonance spectrum of a fiber optic refractive index sensor. *Sensors and Actuators B: Chemical, 156*(2), 559–562.
89. Stewart, D. A., & Léonard, F., (2004). Photocurrents in nanotube junctions. *Physical Review Letters, 93*(10), 107401.
90. Stewart, D. A., & Léonard, F., (2005). Energy conversion efficiency in nanotube optoelectronics. *Nano Letters, 5*(2), 219–222.

91. Takimiya, K., Shinamura, S., Osaka, I., & Miyazaki, E., (2011). Thienoacene-based organic semiconductors. *Advanced Materials, 23*(38), 4347–4370.

92. Tao, Y., Xiao, J., Zheng, C., Zhang, Z., Yan, M., Chen, R., & Huang, W., (2013). Dynamically adaptive characteristics of resonance variation for selectively enhancing electrical performance of organic semiconductors. *Angewandte Chemie, 125*(40), 10685–10689.

93. Tersoff, J., Freitag, M., Tsang, J. C., & Avouris, P., (2005). Device modeling of long-channel nanotube electro-optical emitter. *Applied Physics Letters, 86*(26), 263108.

94. Tokumoto, M., Anzai, H., Ishiguro, T., Saito, G., Kobayashi, H., Kato, R., & Kobayashi, A., (1987). Electrical and magnetic properties of organic semiconductors,(BEDT-TTF) 2X (X= IBr2, IBrCl, and ICl_2). *Synthetic Metals, 19*(1–3), 215–220.

95. Torabi, S., Jahani, F., Van, S. I., Kanimozhi, C., Patil, S., Havenith, R. W., & Koster, L. J. A., (2015). Strategy for enhancing the dielectric constant of organic semiconductors without sacrificing charge carrier mobility and solubility. *Advanced Functional Materials, 25*(1), 150–157.

96. Valentini, L., Mengoni, F., Armentano, I., Kenny, J. M., Ricco, L., Alongi, J., & Mariani, A., (2006). Enhancement of photoelectrical properties in polymer nanocomposites containing modified single-walled carbon nanotubes by conducting dendrimer. *Journal of Applied Physics, 99*(11), 114305.

97. Van Le, Q., Jang, H. W., & Kim, S. Y., (2018). Recent advances toward high-efficiency halide perovskite light-emitting diodes: Review and perspective. *Small Methods, 2*(10), 1700419.

98. Wang, F., Dukovic, G., Brus, L. E., & Heinz, T. F., (2005). The optical resonances in carbon nanotubes arise from excitons. *Science, 308*(5723), 838–841.

99. Wei, J., Zhu, H., Wu, D., & Wei, B., (2004). Carbon nanotube filaments in household light bulbs. *Applied Physics Letters, 84*(24), 4869–4871.

100. Xia, R., Heliotis, G., & Bradley, D. D., (2004). Semiconducting polyfluorenes as materials for solid-state polymer lasers across the visible spectrum. *Synthetic Metals, 140*(2, 3), 117–120.

101. Xia, R., Heliotis, G., Hou, Y., & Bradley, D. D., (2003). Fluorene-based conjugated polymer optical gain media. *Organic Electronics, 4*(2, 3), 165–177.
102. Yamashita, Y., (2009). Organic semiconductors for organic field-effect transistors. *Science and Technology of Advanced Materials, 10*(2), 024313.
103. Yan, M., Rothberg, L. J., Papadimitrakopoulos, F., Galvin, M. E., & Miller, T. M., (1994). Spatially indirect excitons as primary photoexcitations in conjugated polymers. *Physical Review Letters, 72*(7), 1104.
104. Yan, M., Rothberg, L., Hsieh, B. R., & Alfano, R. R., (1994). Exciton formation and decay dynamics in electroluminescent polymers observed by subpicosecond stimulated emission. *Physical Review B, 49*(14), 9419.
105. Yap, B. K., Xia, R., Campoy-Quiles, M., Stavrinou, P. N., & Bradley, D. D., (2008). Simultaneous optimization of charge-carrier mobility and optical gain in semiconducting polymer films. *Nature Materials, 7*(5), 376–380.
106. Zavelani-Rossi, M., Perissinotto, S., Lanzani, G., Salerno, M., & Gigli, G., (2006). Laser dynamics in organic distributed feedback lasers. *Applied Physics Letters, 89*(18), 181105.
107. Zhang, F., Zang, Y., Huang, D., Di, C. A., Gao, X., Sirringhaus, H., & Zhu, D., (2015). Modulated thermoelectric properties of organic semiconductors using field-effect transistors. *Advanced Functional Materials, 25*(20), 3004–3012.
108. Zhang, J., Xu, W., Sheng, P., Zhao, G., & Zhu, D., (2017). Organic donor–acceptor complexes as novel organic semiconductors. *Accounts of Chemical Research, 50*(7), 1654–1662.
109. Zhang, Y., Sun, H., Zhang, S., Li, S., Wang, X., Zhang, X., & Guo, Z., (2019). Enhancing luminescence in all-inorganic perovskite surface plasmon light-emitting diode by incorporating Au-Ag alloy nanoparticle. *Optical Materials, 89*, 563–567.
110. Zhao, H., & Mazumdar, S., (2004). Electron-electron interaction effects on the optical excitations of semiconducting single-walled carbon nanotubes. *Physical Review Letters, 93*(15), 157402.

111. Zhu, Y., Kulkarni, A. P., Wu, P. T., & Jenekhe, S. A., (2008). New ambipolar organic semiconductors. 1. Synthesis, single-crystal structures, redox properties, and photophysics of phenoxazine-based donor– acceptor molecules. *Chemistry of Materials, 20*(13), 4200–4211.

112. Zou, J., & Zhang, Q., (2012). Fundamental properties of carbon nanotubes. *Q. Zhang, Carbon Nanotubes and Their Applications,* 1–30.

CHAPTER 7

ORGANIC SEMICONDUCTORS FOR VISIBLE LIGHTS COMMUNICATORS

CONTENTS

7.1. INTRODUCTION

Organic semiconductors (OSC) are a special kind of polymer product with their own set of unique electrical and optical properties. They are especially interesting because they combine cutting-edge electrical and optical semiconducting properties with easy production and adjustment of organic properties. This makes them stand out because it gives them a unique mix of traits. Unlike organic semiconductors, inorganic semiconductors don't need a highly organized crystalline structure to show their electronic and optical properties. Most of these qualities come from the way semiconductor materials are made chemically. This makes it easier for the company to make devices. For example, they can now be made by depositing from a solution, which is expected to lower the overall cost of both devices. Also, it makes it possible to make many devices at the same time, like the blue, green, and red organic light-emitting diodes (OLEDs) that are used in the screens of smartphones and TVs. (Lin et al., 2012; Godumala et al., 2019).

The optical and electrical properties of OSC derive from their chemical structure, which consists of alternating single and double bonds between adjacent carbon atoms, known as conjugation. This leads to the overlap of electronic orbitals and electron delocalization (Figure 7.1(a)). In conjugated molecules, the carbon atoms are sp^2-hybridized. Three of the four valence electrons are in orbitals in the plane of the molecule which form σ-bonds which hold the molecule together. The fourth valence electron is located in a *p* orbital extending above and below the plane of the molecule. The overlap between the $2pz$ orbitals results in the formation of π-bonds and (when there are many atoms) π and π* bands. Electrons are delocalized along the molecule, and it turns out that there are two such electrons per repeat unit, leading to a filled band and semiconducting electrical and optical properties. This means that they can be used to make a range of semiconducting devices including light-emitting diodes (LEDs), solar cells, transistors, and lasers (Tanaka et al., 2003; Tsonev et al., 2014).

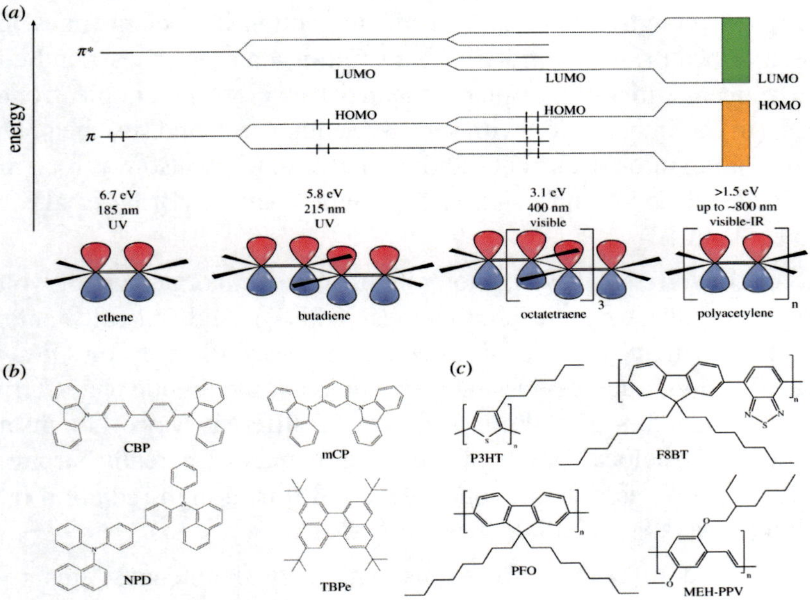

Figure 7.1. (a) A schematic depicting the relationship between the energy levels of alkenes and their conjugation lengths. Because of this, absorption and emission occur at longer wavelengths as the conjugation length increases. This is because the energy gap between the HOMO and LUMO reduces as the conjugation length increases. (b) The chemical structures of some organic small-molecule semiconductors are frequently utilized for organic light-emitting diodes. (c) The chemical structures of some organic polymers are used often as semiconductors.

Source: https://chem.libretexts.org/Bookshelves/Organic_Chemistry/Map%3A_Organic_Chemistry_%28Bruice%29/13%3A_Mass_Spectrometry_Infrared_Spectroscopy_and_Ultraviolet_Visible_Spectroscopy/13.19%3A_The_Effect_of_Conjugation_on_max.

The organic superconductors that are most often utilized originate from one of three categories: small particles, polymeric, or dendrimers. Many are soluble, which makes deposition from solutions a fairly straightforward process; others are evaporated into thin films for use in electronic devices and then applied. Visible region communication is just a relatively recent area wherein OSC were beginning to find many uses (VLCs). Even though the beam is used for interaction since antiquity, optical fibers have only dominated communication systems in the last 30 years, and so much more lately, there has been a great deal of interest in available storage VLC. This is even though light was used for interaction in ancient times. Light has several beautiful qualities for transmitting data, such as hundreds of terahertz license in the first-place channel capacity, (ii) simple front-end devices, (iii) no intervention with sensitive equipment, and (iv) the potential for assimilation into the current lighting facilities. Light also was recognized as Li-Fi, which is an abbreviation for "light fidelity" (Brien et al., 2008; Hanzo et al., 2012).

Color converters, OLEDs, photodiodes (with both organic photovoltaics (OPVs) as well as organic photodiodes (OPDs)), and optically antennas are the four primary types of devices that are based largely on OSC with VLC application. Such devices take advantage of the unique characteristics that superconductors provide in a variety of different ways. For instance, the widespread delocalization of electrons results in a really strong rate of absorption but, as a direct consequence of this, a high radiation rate of emission (Tanaka et al., 2000; Haas et al., 2016).

When used to color conversions with optical antennas, this results in a rapid reaction time when switching on and off. However, the ability to modify the energy deficit by altering the chemical composition helps improve the absorption of OPVs as well as OPDs, as well as the emissions of OLEDs. This is because of the range of both the energy gap. Outcomes in the case of the bandwidth that can be achieved and the data rates that can be achieved are provided throughout this whole chapter. In this analysis, the term "bandwidth" depends on the frequency with which the electromagnetic power output of both the detector drops to a level of–3 dB, which is equivalent to 50% of its value when referenced to DC. Transmission rates also are given, even though they are impacted by a wide variety of parameters including the encoding technique that is used and the possible SNR. When it is in our power to do so, we will offer the most important experimental information linked with the stated data rates (Xie et al., 2002; Sheu et al., 2003).

7.2. ORGANIC SEMICONDUCTORS (OSC) AS COLOR CONVERTERS

The concurrent use of lighting for communications and illuminating is among VLC's goals. Solid-state illumination technologies are a tried-and-true method for offering affordable, high-quality illumination. The white light main source of blue-emitting GaN/InGaN LEDs that have been covered with wavelength-convertible phosphors (Krames et al., 2007; Ahn et al., 2019). The phosphor's function is to absorb several of the LED's blue emissions as well as transform them into yellow emissions, producing white light (Figure 7.2(a)). A wide emitter encompassing the green to red range of the spectrum should be used to provide high-quality lighting, which is assessed by the color rendering index (CRI). Several emitters (green and red phosphors) may produce another even higher CRI.

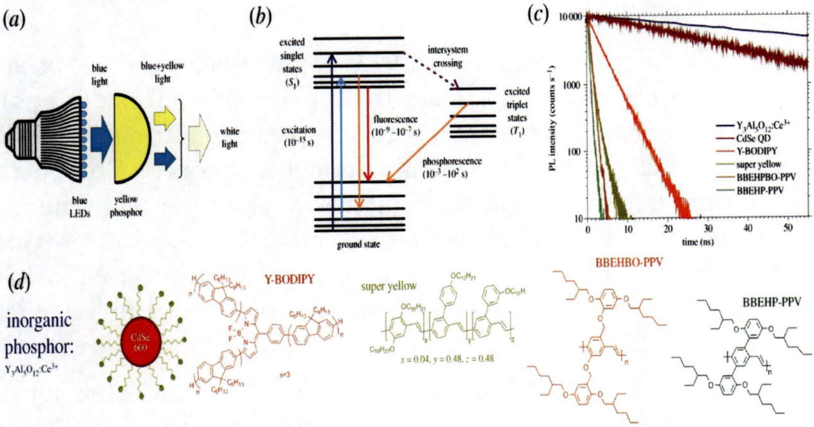

Figure 7.2. The fundamental idea behind a white LED is that uses a phosphor color converter. A phosphor color converter partly absorbs a blue LED's electroluminescence.

Source: https://pubs.acs.org/doi/10.1021/acs.jpcc.2c01679.

Phosphors are excellent at producing white light using LEDs, but they have a delayed reaction time and it's a good bottleneck to VLC systems because of their lengthy illumination lifespan (~ microseconds) (Figure 7.2(b) and (c)). Yellow filtering has been used in some earlier white-light VLC publications to remove the sluggish phosphor emissions and speed up the information transfer. This method has the drawback of using just a fraction of the lighting power for data transfer, which reduces the possible data rate as well as signal-to-noise ratios (SNR) (Xie and Hirosaki, 2007).

To find substitutes for rare-earth phosphor color changers, a new discipline has only lately emerged. Due to the scarcity and un recyclability of conventional rare-earth phosphors, such materials were developed. Families including metal-organic structures, nanoparticles, and even light-emitting polymeric are examples of these alternatives. But a suitable material must meet several requirements to be an efficient color converter for VLC. The first need is that the substance is appropriate for lighting: (I) To accommodate the nitride LEDs used only for illumination, it must meet the following requirements: I must accumulate in the range of 450 nm; (ii) it must have a high photoluminescence quantum yield (PLQY), which would be the amount of emitted photons for every photoelectric effect absorbed; (iii) it must emit within yellow part of the electromagnetic spectrum; (iv) the emitted light must be sufficiently broad to accomplish the high quality of luminance (high CRI).

Additionally, in particular for such VLC applications, (v) the substances' time reaction must be quick (just under 10 ns, preferable just under 1 ns) and similar to the reaction of all the other parts of a data connection. It is difficult to meet all of the aforementioned requirements at once. Consequently, finding the right material might be difficult. We point out that criteria (iii) and (iv) would be adjusted to the mix of emitters utilized to provide an effective color gamut in wavelength-division multiplexed (WDM) VLC situations, where many laser beams of various colors are employed (Smet et al., 2011).

Chun et al. revealed a method for white-light VLC in 2014 that took into account the aforementioned factors and substituted a semiconducting color converter for such phosphor. High radiation rates are necessary to achieve the dual requirements of short emissions lifespan and strong PLQY, which, as was already indicated, is a characteristic of covalently bonded materials. As a result, the scientists produced white light from such an inorganic μ-LED using the readily accessible conjugated polymers "ultra-yellow" (SY). SY's quick time responsiveness (Figure 7.2(c)) made it possible to attain a bandwidth of over 200 MHz. Significantly, such bandwidth was attained without spectrum filtering just at reception and was only constrained by other system elements (the excitation sources and also the detector). They recorded a 1.68 Gbps data throughput even though they only managed to attain an illumination of 240 lx, in part because of the limited maximum output of such micro-LEDs they utilized. Just at the time of publishing, this set records for white-light VLC links and showed that employing quick fluorescence color converters to increase data rate is feasible (Pust et al., 2014).

The work mentioned above was accompanied by additional studies looking at OSC as color converter topologies for VLC applications. Sajjad et al.'s research demonstrated the potential of mixtures of 2 different organic molecules as quick color changes. By adjusting the ratio of two polymeric materials as well as the layer thickness, it offers a practical method of color tuning. Due to the simultaneous emissions of a polymer, any CIE co-ordinates are attainable in an InGaN LED/polymer 1/polymer 2 while retaining strong white light quality (i.e., a high premium of CRI). In just this work, a program called Förster resonance (FRET) that occurs on a picosecond timeframe excites one polymer as well as transfers some of its power generations to the other. The power transmission mechanism, which adds another stimulation decay route and accelerates the fluorescence of the first activated polymer, makes this fascinating. For instance, it was shown in publishing (Hirosaki, 2007) that now the lifespan of the BBEHP-PPV (τ =0.83 ns) may be lowered to 0.27 ns when combined with MEH-PPV (τ =0.38 ns). The group was able to accomplish large bandwidth (more than 200 MHz), acceptable CRI, and transfer speeds as a result (greater than 350 Mbps) (Chun et al., 2014).

As color conversions for VLC, several articles have looked at specifically produced organic and polymeric substances. Star-shaped compounds that contain a BODIPY core with various side arm lengths were examined by Vithanage et al. They showed that now the greatest data rates of 370 Mbps, as well as the largest bandwidth of 73 MHz, are achieved by Y-BODIPY particles (Figure 7.2(d)) with three fluorene subunits per arm (Y-B3). Figure 7.2(c) depicts the material's moment luminescence. The fluorescent decay of Y-B3 is substantially quicker than that of II-VI quantum dots or a conventional phosphor (CL-827), however, the conjugated polymers are already faster, which makes them more desirable. BBEHP-PPV is a greenish emission material, yet it has a very quick fluorescence decline. As a result, BBEHBO-PPV, a novel linked polymer, was created with redshifted emissions. The BBEHP-bulky PPV's poly(2,5-dialkoxy-p-phenylene vinylene) backbone was preserved; however, the terminal groups were connected via an alkoxy bridging (as in MEH-PPV). The substance demonstrated a high frequency of 470 MHz while reaching a data throughput of 550 Mbps using on-off keying due to its relatively brief fluorescence lifespan (0.37 ns). AIEgens, or accumulation emissions luminous, have been used as color changes, according to Zhang et al. The entire properties of AIEgens, ranging from blue to red, have already been studied by the authors (Khalid et al., 2012; Wang et al., 2017). These substances have a high electrical bandwidth capacity of up to 279 MHz 6 Db (about 110 MHz 3 dB), which enables them to reach data speeds of over 500 Mbps (Sun et al., 2003). It's interesting to note that the researchers also empirically supported the possible relation between both the modulating frequency and the decrease of the photoluminescence (PL) entire life:

$$BW_{-6dB} = \sqrt{3} / (2\pi\tau)$$

Other VLC color conversion research has focused on perovskite nanoparticle utilization and methods to improve the behavior of organic solar cells using metamaterials (Kim et al., 2010; Cui et al., 2015). Utilizing nanopatterned hyperbolic meta surface to boost the radiation velocity of the semiconducting color conversion, Yang et al. made use of the Purcell phenomena. In particular, the scientists increased the super yellow polymer's frequency by 67% while reducing the intensity of emissions. In practical terms, this made it possible to raise the transmission data rate again for the nanopatterned devices from 250 Mbps with the frequency of 125 MHz for planar super yellow films to 150 Mbps and a frequency of 75 MHz for flat super yellow films. Perovskite quantum dots have been investigated by Mei et al. as a color converter. Their 300 Mbps data transmission and 73 MHz bandwidth were shown using CsPbBr1.8I1.2 quantum dots that produce yellow light (Stupca et al., 2012; Song and Yang, 2012).

When used as color changers for blue-emitting LEDs, OSC are a desirable replacement for inorganic phosphors. Their major benefit is their quick time reaction, but they also have the advantage of being able to adjust the light output for customized light sources. Their short emitting lifespan (sub-nanosecond to several nanoseconds) gets around the standard phosphors' sluggish reaction time (millisecond range) bottleneck. Could not only do superconductors construct more effective white-light VLC connections, and so they can also create higher light inexpensively and independently of the rare minerals used in inorganic phosphors, which gives them significant promise as a color converter in illumination (Jang et al., 2013; Chen et al., 2014).

7.3. ORGANIC LIGHT-EMITTING DIODES (OLEDS) AS LIGHT SOURCES

Organic light-emitting diodes (OLEDs), photovoltaics, and transistors are just a few examples of the semi-conductive components that may be made using OSC. One or even more organic light-emitting layers sandwiched among appropriate connections make up an OLED. A voltage causes it to produce light. Although Pope and colleagues first described electroluminescence (EL) in such singular anthraquinone crystal in 1963, Tang and VanSlyke and Burroughes et al. are widely credited with developing the first practical thin layer OLEDs. OLEDs were first created to be used in screens, and now they are found in many professional mobile phone screens as well as certain TVs (Heliotis et al., 2005; Zhou et al., 2015).

They are presently being explored for illumination, and smart devices, including wearables because of their distinctive manufacturing and emissions features. Due to the advancements achieved, OLEDs may now be used as broadcasters for VLC, and this field of study has recently gained a lot of attention.

Figures 7.3(a)–(c) depict the operation of an OLED with the most straightforward construction. Among both connections is an organic semiconductor light-emitting film. Indium tin oxide (ITO) is often employed because one of the connections has to be translucent, and its tremendous work performance makes it suited to serve as the anode, that is, to introduce pores into the organic phase. The organic layer receives electrons from the opposite electrode (cathode), metals. Underneath the influence of the electric field, the inserted ions and electrons migrate over the organic phase, and charged couples bond to produce excitons (Figure 7.3(b)). The exciton may then undergo thermal decay, which allows some light to pass through to the translucent contact and leave the device. Because electron/hole infusion and transportation are often out of equilibrium, such a simple structure's effectiveness is typically poor. Consequently, to manage charge injections, transportation, and combination and achieve significantly greater performance, modern OLEDs often comprise extra layers (Figure 7.3(d)). By reducing the energy difference for injecting from the electrodes, the inclusion of extra layers for the injecting of the depletion layer addresses the issue of unbalanced carriers. Furthermore, carriers may be prevented from exiting the particle emitters layer without coupling by hole or electron carrier layers. In, the operating principles of OLEDs are described in further detail (Geffroy et al., 2006; Gu et al., 2007).

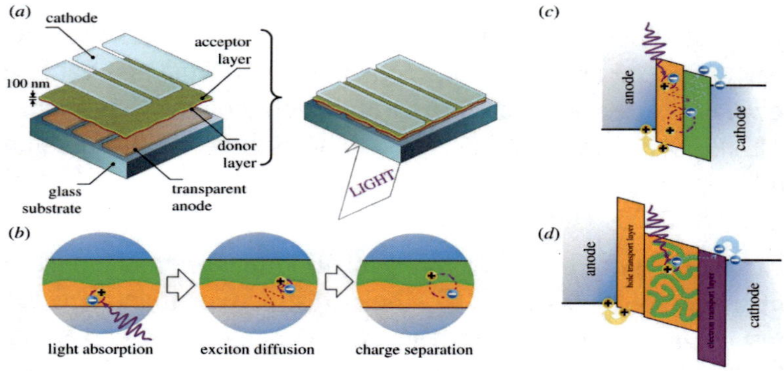

Figure 7.3. The fundamentals of OLED functioning.

Source: https://www.luna-electronic.de/OLED-Produktinformationen/.

The attraction of OLEDs is due to their easy manufacture, emission properties, flexibility, and ability to be incorporated into some other things, as was before indicated. While certain OLED substances are thermally vaporized, others may be formed from solutions. This removes the requirement for high-temperature epitaxial, enables straightforward patterning of several colors next to one another, and lowers the total cost of manufacturing. Curved displays and wearable technology are now possible because of the usage of flexible substrates made possible by the OLED manufacturing process. The particle emitters surface of OLEDs may be made of a virtually infinite wide range of chemical molecules, making it feasible to create and adjust features like emission color and color quality. Only lately has interest grown in the potential the use OLEDs as VLC transmitters.

Due to the limited motility of semiconductor materials as well as the high capacitor of these extremely thin devices, which often results in a really limited bandwidth, these aren't obvious options for communications. When compared to the inorganic compounds used during LEDs (about 1,000 cm^2 $(V\ s)^{-1}$), the usual charge vibrations observed for OLEDs are several orders of magnitude smaller (10^{-3}–10 cm^2 $(V\ s)^{-1}$) (Findlay et al., 2014). With OLEDs, this is somewhat offset by the fact that the devices are significantly thinner (tens to hundreds of nanometers). The bandwidth of these big OLED pixels (more than mm^2) is severely constrained by this thin device design's high capacitances, which are caused by this thin device structure (Findlay et al., 2014; Sajjad et al., 2015, p. 17).

Minh et al.'s work is among the first to employ an OLED as a VLC broadcaster. They quoted a 2.15 Mbps data rate and a 150 kHz frequency. The scientists used an efficient equalization like one tactic to reduce the impact of the OLED's high capacitance. Despite the massive OLED surface (3.4 cm^2, Lumiblade OLED by Philips) and therefore high capacitance of a device (about 0.3 F), this allowed a VLC connection to function at multiple Mbps. As a consequence, the scientists increased the device size's attainable frequency and showed that 1 MHz could be achieved. Another commercialized OLED (Osram ORBEOS CMW-031) used by the same outfit was also claimed to support VLC. The very same equalization technique was used, and they recorded a data rate of 550 kbps, which is nine times greater than it would have been without normalization (Vithanage et al., 2016, p. 17).

The very same team produced a series of articles in 2013 that investigated various modulation strategies and broke the old best for VLC data transmission using OLEDs. Haigh et al. employed a discrete multiton—DMT—spectrally effective modulation technique to obtain a data rate of 1.4 Mbps, which was 14 times the frequency of systems used. They obtained a data throughput of 2.7 Mbps and made the first ever online filtering. This early success made OLEDs transmitter for VLC possible, and this field of study has since grown active (Leitao et al., 2017).

Figure 7.4 displays the development of observed data rates. Whenever a data rate of 10 Mbps was attained in 2014, a new milestone was established. Using a specially manufactured OLED, the bit rate record has been further quadrupled that year. The fastest recorded download speeds peaked in 2017 at 51.6 Mbps after reaching 27.9 Mbps in 2015. When Yoshida et al. revealed OLED devices that could attain data speeds over 1 Gbps in a connection 2 m long, a breakthrough was made in 2020. The right selection of emission materials, tiny circuit sizes, carefully engineered connections with low impedance, and effective heat regulation by constructing the transistors on oxidized silicon wafers allowed them to achieve a 20-fold increase in data rate. They were able to run the devices at a voltage level and attain a large bandwidth capacity, exceeding 245 MHz, thanks to efficient heat management (Zhang et al., 2018).

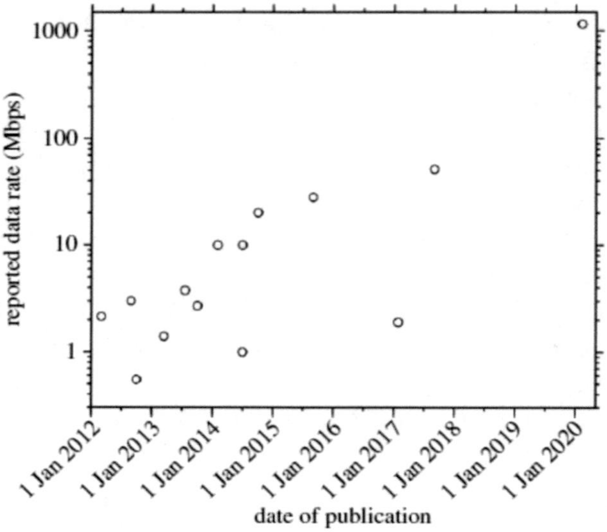

Figure 7.4. *OLED device reported VLC data rates as a function of publication date.*

Source: https://ietresearch.onlinelibrary.wiley.com/doi/full/10.1049/iet-opt.2019.0120.

A rapidly expanding sector is VLC employing OLEDs. OLEDs are gaining popularity because of their unique qualities as contrasted to inorganic LEDs, such as flexible substrates, scalable manufacture of several emitting colors side-by-side, adjustability, and inexpensive production costs. Scientists have shown that the data speeds achieved have improved quickly, despite the restricted capacity of OLEDs.

7.4. ORGANIC PHOTODIODES AND PHOTOVOLTAICS

When the recipient or broadcaster is just stand-alone equipment without a direct connection to an electricity network, this is a common VLC situation. Monitoring devices, portable home equipment and gadgets, including mobile virtual assistants, household robotics, and smartphones are instances of this situation. Using photovoltaics as just a source of power and so as VLC signal receivers is just an intriguing solution to the issue of powering such devices. The concept would be for the solar cell to perform a dual role, converting light as well as ambient energy for communications into electric energy for a standalone unit (Laurand et al., 2012). Figure 7.5 depicts the OPVs' basic operating concept.

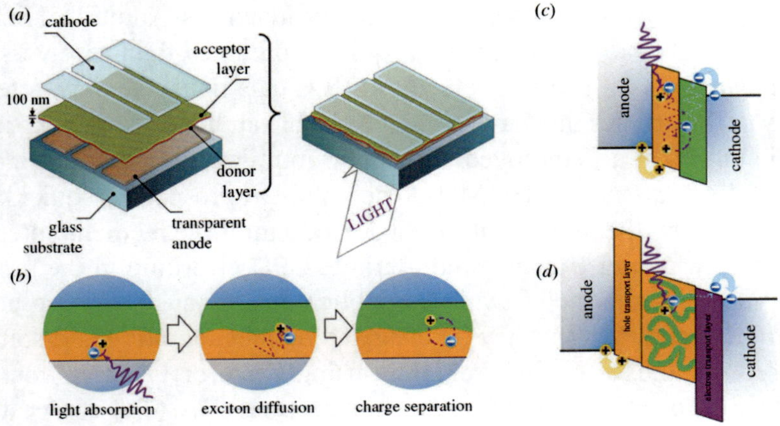

Figure 7.5. The fundamental idea behind how an OPV cell works.

Source: https://royalsocietypublishing.org/doi/10.1098/rsta.2019.0186.

One or more semiconducting layers sandwiched between electrical contacts, one being transparent, make up the fundamental geometry of an OPV, which is comparable to those of OLEDs (Figure 7.5(a)). The organic molecules absorb visible light, and so as a consequence, a coupled electron-hole pair, or exciton, is produced (Figure 7.5(b)). The differential in energy state between both the materials separates the charges when the excitons migrate to contact among donor-acceptor materials (or areas of the sample) (Figure 7.5(c)). Just at contacts, the separate electrons are finally removed. In so many contemporary OPVs, donating and receiver substances are combined to create what is known as just a bulk heterojunction, whereas extra layers permit the transportation of the necessary charges but prevent the opposing ions from accessing the "wrong" contact (Figure 7.5(d)) (Yang et al., 2018).

In 2013, the very first reports on the use of OPDs as VLC sensors were released. If organic heterojunction OPVs built on P3HT: PCBM can function as photodetectors in VLC, Arredondo et al. looked into that possibility. They showed a modulating frequency of 790 kHz while reverse biasing the OPV. We see that the term "reverse biased" describes an OPV that functions as a photodiode but does not produce power. A flexible, all-organic VLC system was presented by Vega-Colorado et al. Flexible OPVs were created in this work using a roll-to-roll method and served as a VLC receiver. To increase the frequency up to 200 kHz, they additionally used the OPV with reverse bias. The authors of the study used audio data transfer to illustrate the potential of a VLC connection composed solely of organic materials. Additional findings show that OPDs may be employed as quick modulated light receivers even when they are not for use as a detector in VLC data lines. We draw attention to a few of the articles that have attracted sizable audiences. In an intriguing study, Peumans et al. employed alternating multilayered photodetectors to reach a 3 dB frequency of 430 MHz. Their ability to respond so quickly was made possible by the thin multilayered device's multiple recipient interfaces and relatively shorter excited-state lifetime. OPDs working in the range of 50–80 MHz was reported in various publications by such a group headed by Ohmori (Tang, 1987). They investigated both evaporation and solution-processed complexes in their research, using CuPc or F8T2 as donors as well as BPPC or PCBM as recipients, if appropriate. Last but not least, Tsai et al. claimed that photodiodes were able to achieve 30 MHz frequency by using the large vertical carrier concentration. Pentacene served as the donor material, while C_{60} served as the given input.

Zhang et al.'s utilization of the OPV in 2015 was more creative (Burroughes et al., 1990; Choi et al., 2017). Instead of reverse biasing the OPV in just this research, the authors modified a transceiver (Figure 7.6) to collect illumination power while still receiving network information. The reception circuit's energy collection and data transfer branches included low- and high-frequency filtering, enabling the OPV to perform both functions simultaneously. The researchers discovered that this setup could concurrently receive data rates of 34.2 Mbps and harvest 0.43 mW of electricity (5.4 mWcm^2). The recipient's data transfer optimization increased the bit rate to 42.3 Mbps.

There have been very few reports on the use of OPVs as VLC detectors up to this point. They provide the intriguing option of inexpensive and identity solutions for VLC listeners, and hence they have a lot of promise for future VLC applications despite the very limited bitrates and data speeds recorded so far.

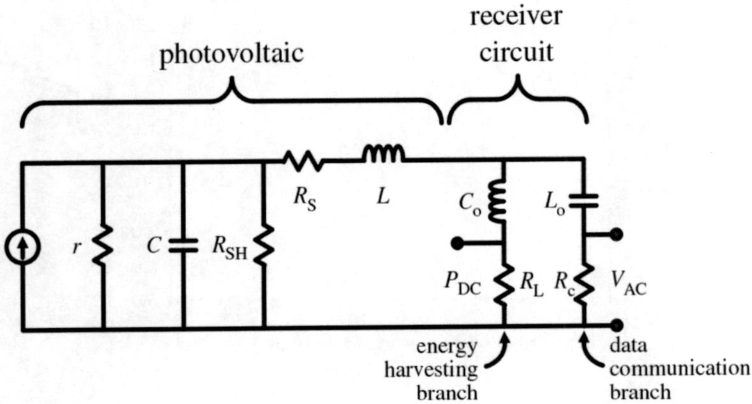

Figure 7.6. Equivalent electric circuit of a solar cell (left portion) and receiver circuit (right part).

Source: https://www.researchgate.net/figure/Equivalent-circuit-of-a-solar-cell-consisting-of-discrete-electrical-components_fig2_318324604.

7.5. FLUORESCENT ANTENNAS FOR VISIBLE LIGHT COMMUNICATIONS (VLCS)

Having sensors which are both quick (high frequency) and responsive (high SNR) is a difficulty for VLC (Figure 7.7(a)). A tiny photodetector does have a broad bandwidth but just a low SNR, making it quick yet insensitive (Figure 7.7(b)). A big photodiode does have a high SNR but a limited bandwidth, making it sensitive yet sluggish (Figure 7.7(c)). This may be avoided by using a lens or another optical component, such as a compound parabolic extractor, which collects light across a larger area and concentrates it into a smaller region. Consequently, it may boost a tiny photodetector's SNR (Figure 7.7(d)). To obtain the necessary SNR and capacity for high-speed communication, the majority of VLC systems depend on small focal length lenses to concentrate the light into quick photodiode detectors. However, in situations in which the receivers remain mobile, this setup requires precise alignments to the sources (Figure 7.7(d)), which is not feasible (e.g., a laptop or a smartphone). Recent research has shown that a different strategy can get around this restriction and provide a sizable gain with a broad field of view by employing photonic systems that rely on fluorescent and attenuated total reflections (Carey et al., 2017; Park et al., 2018).

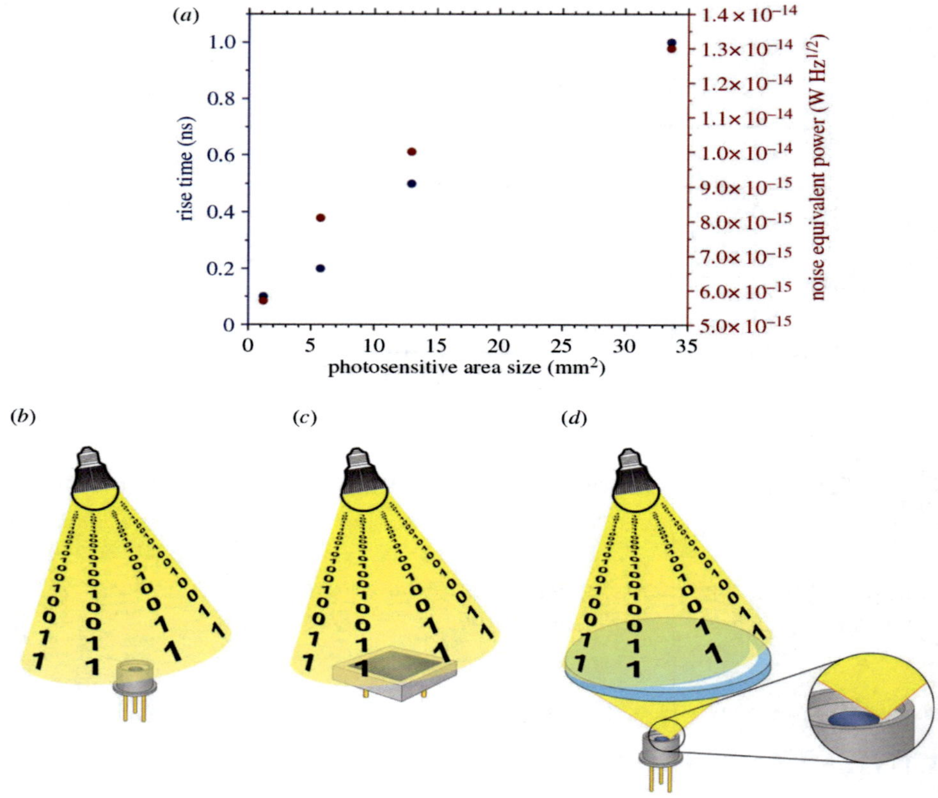

Figure 7.7. Plot showing the rising time as well as noise comparable power concerning silicon photodiode photosensitive region.

Source: https://www.sciencedirect.com/topics/engineering/photodetector.

For just any electro-optic system that relies on reflection and refraction, there would be a trade-off between both the photonic benefit (the power output frequency, Sout, divided either by feedback density, Sin) as well as the ground of structure of a system (in) due to a basic aspect in optics called the preservation of étendue. Practically speaking, this implies that now the field of vision of an objective lens is limited by how much gain it can produce (Figure 7.8(b)). Because of this, although the VLC system's means of optical components may attain greater SNR because of the good optical amplification, they often have a relatively small field of vision (Bansal et al., 2015).

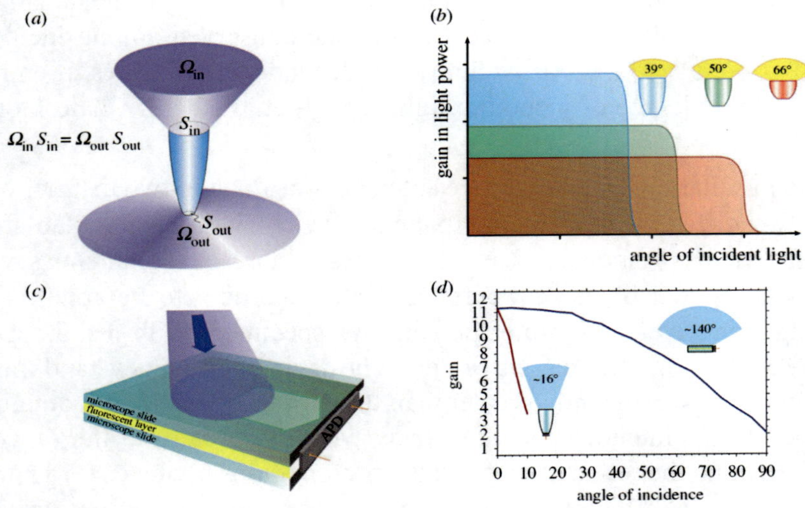

Figure 7.8. (a) preservation of expanse. (b) Gain and field of vision for three compound parabolic concentrators are compared graphically. (c) The fundamental workings of a fluorescent antenna. (d) Gain versus angle for a fluorescent antenna and a focusing lens are experimentally compared. A lens with a comparable gain has a field of vision of around 2×8°, but the fluorescent antenna has a field of view (full width at half maximum) of roughly 2×70°.

Source: https://www.sciencedirect.com/science/article/abs/pii/S0360544220320399.

To get around the restrictions for optical systems imposed either by conserving of étendue, Collins et al. hypothetically showed in 2014 that a fluorescence antenna, a device comparable to a luminous solar concentrator, may be utilized in VLC. These devices might retain a large field of vision while offering gain comparable to that of traditional optics. Two different groups independently produced practical proof of this idea in 2016. Figure 7.8(c) demonstrates how these devices work: incoming light that is absorbed in such a coating of the fluorescent lamp before being re-emitted emits fluorescence. Total internal reflection traps the majority of both the light emitted inside the layer, where it travels to the satellite's edge within the instrument. Significant optically gain is possible due to such a device's edge having a significantly smaller cross-section than that of the absorption region, and the field of vision is governed by the cosine law. Figure 7.8(d)'s practical result illustrates that all of these devices may provide a substantially broader field of vision than optics with comparable optical improvements (Figure 7.8(d)). This results from the lack of étendue conserved among incidents as well as generated light as well as the independence of angles of emitted light from the degree of incoming light (Haigh et al., 2013; Chaleshtori et al., 2018).

The installation of fluorescence antennas was done in two distinct ways in the two different experimental papers. Devices with such a slab shape were shown by Manousiadis et al. whereas fluorescence antennas were built using optical fibers by Peyronel et al. According to Peyronel et al., the gadget was created by interspersing two specimens with just an epoxy mixture containing a fluorescent protein. The devices' frequency band was 40 MHz, which is significantly greater than those of the majority of commonly produced LEDs intended for luminance, which is usually in the range of 5–20 MHz, and the researchers proved representations of objects of 12, with entire width at half highest visual field of 120°. They also accomplished a bit rate of 190 Mbps, which has been constrained by the recipient used. A bundle of commercialized fluorescent fibers arranged in such a planar array was employed by the researchers.

They achieved a data throughput of 2.36 Gbps while demonstrating an optical gain of 12.3 and a frequency of 91 MHz. Interestingly, the authors presented a design for an omnidirectional antenna using fluorescence fibers. Because the signal could be detected anywhere at the incidence angle, this is significant since it addresses one of the obstacles within actual mobile detectors' VLC applications.

REFERENCES

1. Ahn, Y. N., Kim, K. D., Anoop, G., Kim, G. S., & Yoo, J. S., (2019). Design of highly efficient phosphor converted white light-emitting diodes with color rendering indices (R1−R15) ≥95 for artificial lighting. *Sci. Rep., 9*, 16848. doi: 10.1038/s41598-019-53269-0.
2. Bansal, A. K., Hou, S., Kulyk, O., Bowman, E. M., & Samuel, I. D. W., (2015). Wearable organic optoelectronic sensors for medicine. *Adv. Mater., 27*, 7638–7644. doi: 10.1002/adma.201403560.
3. Brien, D. C. O., Zeng, L., Le-Minh, H., Faulkner, G., Walewski, J. W., & Randel, S., (2008). Visible light communications: Challenges and possibilities. In: *2008 IEEE 19th Int. Symp. on Personal, Indoor and Mobile Radio Communications* (pp. 1–5). Cannes, France. Piscataway, NJ: IEEE. doi: 10.1109/PIMRC.2008.4699964.
4. Burroughes, J. H., Bradley, D. D. C., Brown, A. R., Marks, R. N., Mackay, K., Friend, R. H., Burns, P. L., & Holmes, A. B., (1990). Light-emitting diodes based on conjugated polymers. *Nature, 347*, 539–541. doi: 10.1038/347539a0.
5. Carey, T., et al., (2017). Fully inkjet-printed two-dimensional material field-effect heterojunctions for wearable and textile electronics. *Nat. Commun., 8*, 1202. doi: 10.1038/s41467–017-01210-2.
6. Chaleshtori, Z. N., Chvojka, P., Zvanovec, S., Ghassemlooy, Z., & Haigh, P. A., (2018). A survey on recent advances in organic visible light communications. In: *2018 11th Int. Symp. on Communication Systems, Networks & Digital Signal Processing (CSNDSP), Budapest, Hungary* (pp. 1–6). Piscataway, NJ: IEEE. (doi: 10.1109/CSNDSP.2018.8471788.
7. Chen, J., Liu, W., Mao, L. H., Yin, Y. J., Wang, C. F., & Chen, S., (2014). Synthesis of silica-based carbon dot/nanocrystal hybrids toward white LEDs. *J. Mater. Sci., 49*, 7391–7398. doi: 10.1007/s10853-014-8413-y.
8. Choi, S., Kwon, S., Kim, H., Kim, W., Kwon, J. H., Lim, M. S., Lee, H. S., & Choi, K. C., (2017). Highly flexible and efficient fabric-based organic light-emitting devices for clothing-shaped wearable displays. *Sci. Rep., 7*, 6424. doi: 10.1038/s41598-017-06733-8.

9. Chun, H., et al., (2014). Visible light communication using a blue GaN μLED and fluorescent polymer color converter. *IEEE Photon. Technol. Lett., 26*, 2035–2038. doi: 10.1109/LPT. 2014.2345256.
10. Cui, Y., Song, T., Yu, J., Yang, Y.,Wang, Z., & Qian, G., (2015). Dye encapsulatedmetal-organic framework for warm-white LED with high color-rendering index. *Adv. Funct. Mater., 25*, 4796–4802. doi: 10.1002/adfm.201501756.
11. Findlay, N. J., Bruckbauer, J., Inigo, A. R., Breig, B., Arumugam, S., Wallis, D. J., Martin, R. W., & Skabara, P. J., (2014). An organic down-converting material for white-light emission from hybrid LEDs. *Adv. Mater., 26*, 7290–7294. doi: 10.1002/adma.201402661.
12. Geffroy, B., Le Roy, P., & Prat, C., (2006). Organic light-emitting diode (OLED) technology: Materials, devices and display technologies. *Polym. Int., 55*, 572–582. doi: 10.1002/pi.1974.
13. Godumala, M., Choi, S., Cho, M. J., & Choi, D. H., (2019). Recent breakthroughs in thermally activated delayed fluorescence organic light emitting diodes containing non-doped emitting layers. *J. Mater. Chem. C, 7*, 2172–2198. doi: 10.1039/C8TC06293E.
14. Gu, E., Zhang, H. X., Sun, H. D., Dawson, M. D., Mackintosh, A. R., Kuehne, A. J. C., Pethrick, R. A., et al., (2007). Hybrid inorganic/organic microstructured light-emitting diodes produced using photocurable polymer blends. *Appl. Phys. Lett., 90*, 031116. doi: 10.1063/1.2431772.
15. Haas, H., Yin, L., Wang, Y., & Chen, C., (2016). What is LiFi? *J. Lightwave Technol., 34*, 1533–1544. doi: 10.1109/JLT.2015.2510021.
16. Haigh, P. A., Ghassemlooy, Z., Rajbhandari, S., & Papakonstantinou, I., (2013). Visible light communications using organic light emitting diodes. *IEEE Commun. Mag., 51*, 148–154. doi: 10.1109/MCOM.2013.6576353.
17. Hanzo, L., Haas, H., Imre, S., Brien, D. O., Rupp, M., & Gyongyosi, L., (2012). Wireless myths, realities, and futures: From 3G/4G to optical and quantum wireless. *Proc. IEEE, 100*, 1853–1888. doi: 10.1109/JPROC.2012.2189788.
18. Heliotis, G., Itskos, G., Murray, R., Dawson, M. D., Watson, I. M., & Bradley, D. D. C., (2006). Hybrid inorganic/organic semiconductor heterostructures with efficient non-radiative energy transfer. *Adv. Mater., 18*, 334–338. doi: 10.1002/adma.200501949.

19. Heliotis, G., Stavrinou, P. N., Bradley, D. D. C., Gu, E., Griffin, C., Jeon, C. W., & Dawson, M. D., (2005). Spectral conversion of InGaN ultraviolet microarray light-emitting diodes using fluorene-based red-, green-, blue-, and white-light-emitting polymer overlayer films. *Appl. Phys. Lett., 87*, 103505. doi: 10.1063/1.2039991.

20. Jang, E. P., Song, W. S., Lee, K. H., & Yang, H., (2013). Preparation of a photo-degradation-resistant quantum dot–polymer composite plate for use in the fabrication of a high-stability whitelight- emitting diode. *Nanotechnology, 24*, 045607. doi: 10.1088/0957-4484/24/4/045607.

21. Khalid, A. M., Cossu, G., Corsini, R., Choudhury, P., & Ciaramella, E., (2012). 1-Gb/s transmission over a phosphorescent white LED by using rate-adaptive discrete multitone modulation. *IEEE Photon. J., 4*, 1465–1473. doi: 10.1109/JPHOT.2012.2210397.

22. Kim, O. H., Ha, S. W., Kim, J. I., & Lee, J. K., (2010). Excellent photostability of phosphorescent nanoparticles and their application as a color converter in light emitting diodes. *ACS Nano*, *4*, 3397–3405. doi: 10.1021/nn100139e.

23. Krames, M. R., Shchekin, O. B., Mueller-Mach, R., Mueller, G. O., Zhou, L., Harbers, G., & Craford, M. G., (2007). Status and future of high-power light-emitting diodes for solid-state lighting. *J. Display Technol., 3*, 160–175. doi: 10.1109/JDT.2007.895339.

24. Laurand, N., et al., *(*2012*)*. Colloidal quantum dot nanocomposites for visible wavelength conversion of modulated optical signals. *Opt. Mater. Express, 2*, 250–260. doi: 10.1364/ OME.2.000250.

25. Leitao, M. F., Santos, J. M. M., Guilhabert, B., Watson, S., Kelly, A. E., Islim, M. S., Haas, H., et al., (2017). Gb/s visible light communications with colloidal quantum dot color converters. *IEEE J. Sel. Top. Quantum Electron., 23*, 1–10. doi: 10.1109/JSTQE.2017.2690833.

26. Lin, C. C., & Liu, R. S., (2011). Advances in phosphors for light-emitting diodes. *J. Phys. Chem. Lett., 2*, 1268–1277. doi: 10.1021/jz2002452.

27. Lin, Y., Li, Y., & Zhan, X., (2012). Small molecule semiconductors for high-efficiency organic photovoltaics. *Chem. Soc. Rev., 41*, 4245–4272. doi: 10.1039/C2CS15313K.

28. Mei, S., Liu, X., Zhang, W., Liu, R., Zheng, L., Guo, R., & Tian, P., (2018). High-bandwidth white-light system combining a micro-LED

with Perovskite quantum dots for visible light communication. *ACS Appl. Mater. Interfaces, 10*, 5641–5648. doi: 10.1021/acsami.7b17810.

29. Park, S., et al., (2018). Self-powered ultra-flexible electronics via nano-grating-patterned organic photovoltaics. *Nature, 561*, 516–521. doi: 10.1038/s41586-018-0536-x.
30. Pust, P., et al., (2014). Narrow-band red-emitting SrLiAl3N4]:Eu2+ as a next-generation LEDphosphor material. *Nat. Mater., 13*, 891–896. doi: 10.1038/nmat4012.
31. Sajjad, M. T., et al., (2015). Novel fast color-converter for visible light communication using a blend of conjugated polymers. *ACS Photonics, 2*, 194–199. doi: 10.1021/ph500451y.
32. Sajjad, M. T., et al., (2017). A saturated red color converter for visible light communication using a blend of star-shaped organic semiconductors. *Appl. Phys. Lett., 110*, 013302. doi: 10.1063/1.4971823.
33. Sheu, J. K., et al., (2003). White-light emission from near UV InGaN-GaN LED chip precoated with blue/green/red phosphors. *IEEE Photon. Technol. Lett., 15*, 18–20. doi: 10.1109/ LPT.2002.805852.
34. Smet, P. F., Parmentier, A. B., & Poelman, D., (2011). Selecting conversion phosphors for white light emitting diodes. *J. Electrochem. Soc., 158*, R37–R54. doi: 10.1149/1.3568524.
35. Song, W. S., & Yang, H., (2012). Efficient white-light-emitting diodes fabricated from highly fluorescent copper indium sulfide core/shell quantum dots. *Chem. Mater., 24*, 1961–1967. doi: 10.1021/cm300837z.
36. Stupca, M., Nayfeh, O. M., Hoang, T., Nayfeh, M. H., Alhreish, B., Boparai, J., AlDwayyan, A., & AlSalhi, M., (2012). Silicon nanoparticle-ZnS nanophosphors for ultraviolet-based white light emitting diode. *J. Appl. Phys., 112*, 074313. doi: 10.1063/1.4754449.
37. Sun, C. Y., et al., (2013). Efficient and tunable white-light emission of metal–organic frameworks by iridium-complex encapsulation. *Nat. Commun., 4*, 2717. doi: 10.1038/ncomms3717.
38. Tanaka, Y., Haruyama, S., & Nakagawa, M., (2000). Wireless optical transmissions with white colored LED for wireless home links. In: *11th IEEE Int. Symp. on Personal Indoor and Mobile Radio Communications. PIMRC 2000. Proc. (Cat. No.00TH8525), London, UK* (Vol. 1322, pp. 1325–1329). Piscataway, NJ: IEEE. doi: 10.1109/ PIMRC2000.881634.

39. Tanaka, Y., Komine, T., Haruyama, S., & Nakagawa, M., (2003). Indoor visible light data transmission system utilizing white LED lights. *IEICE Trans. Commun., 86*, 2440–2454.
40. Tang, C. W., & VanSlyke, S. A., (1987). Organic electroluminescent diodes. *Appl. Phys. Lett., 51*, 913–915. doi: 10.1063/1.98799.
41. Tsonev, D., et al., *(*2014*)*. A 3-Gb/s single-LED OFDM-based wireless VLC link using a gallium nitride μLED. *IEEE Photonics Technol. Lett., 26*, 637–640. doi: 10.1109/LPT.2013.2297621.
42. Vithanage, A. D., et al., *(*2016*)*. BODIPY star-shaped molecules as solid state color converters for visible light communications. *Appl. Phys. Lett., 109*, 013302. doi: 10.1063/1.4953789.
43. Vithanage, D. A., et al., *(*2017*)*. Polymer color converter with very high modulation bandwidth for visible light communications. *J. Mater. Chem. C, 5*, 8916–8920. doi: 10.1039/C7TC03787B.
44. Wang, Z., Chen, B., & Rogach, A. L., (2017). Synthesis, optical properties and applications of light emitting copper nanoclusters. *Nanoscale Horiz., 2*, 135–146. doi: 10.1039/C7NH00013H.
45. Xie, R. J., & Hirosaki, N., (2007). Silicon-based oxynitride and nitride phosphors for white LEDs—A review. *Sci. Technol. Adv. Mater., 8*, 588–600. doi: 10.1016/j.stam.2007.08.005.
46. Xie, R. J., Mitomo, M., Uheda, K., Xu, F. F., & Akimune, Y., (2002). Preparation and luminescence spectra of calcium- and rare-earth (R=Eu, Tb, and Pr)-codoped α-SiAlON ceramics. *J. Am. Ceram. Soc., 85*, 1229–1234. doi: 10.1111/j.1151-2916.2002.tb00250.x.
47. Yang, X., Shi, M., Yu, Y., Xie, Y., Liang, R., Ou, Q., Chi, N., & Zhang, S., (2018). Enhancing communication bandwidths of organic color converters using nanopatterned hyperbolic metamaterials. *J. Light. Technol., 36*, 1862–1867. doi: 10.1109/JLT.2018.2793217.
48. Zhang, Y., et al., *(*2018*)*. Aggregation-induced emission luminogens as color converters for visible-light communication. *ACS Appl. Mater. Interfaces, 10*, 34418–34426. doi: 10.1021/acsami. 8b05950.
49. Zhou, D., Zou, H., Liu, M., Zhang, K., Sheng, Y., Cui, J., Zhang, H., & Yang, B., (2015). Surface ligand dynamics-guided preparation of quantum dots–cellulose composites for light-emitting diodes. *ACS Appl. Mater. Interfaces, 7*, 15 830–15 839. doi: 10.1021/acsami.5b03004.

CHAPTER 8

PLASMONICS FOR LIGHT-EMITTING AND PHOTOVOLTAIC DEVICES

CONTENTS

8.1. INTRODUCTION

Surface plasmon (SP), also known as the quantization of the plasma oscillations of unpaired electrons just at metal/dielectric interaction, can pair to electromagnetic field and generates a particular mode just at interface known as surface plasmon polariton (SPP). SPP has the unique capacity to concentrate and intensify magnetic waves, in addition to bringing new optical features and functionalities to materials. It also has the potential to improve material performance. Plasmonics is the term used to describe the method of regulating and exploiting SPP, and this has garnered a lot of interest as a result of the recent fast advancement of nanotechnology (Barnes et al., 2003; Atwater, 2007).

8.2. OPTICAL PROPERTIES OF THE SURFACE PLASMON (SP) RESONANCE

The SPP has two different kinds of phases: propagating forms and confined forms. Figure 8.1(a) presents the distribution of an electromagnetic current for the SPP propagation mode as determined by three-dimensional (3D) finite amount of time (3D-FDTD) calculations. The arrows represent the directions of electrostatic charge produced either by plasma oscillation variations within the electronic structure. Only the metal surface is affected by the electrostatic potential compressional waves (Nakamura et al., 1993). The SPP's transverse wave of charged fluctuations just at the surface of the metal allows it to interface with light waves, whereas the SP's collisional wave had only a longitudinally wave. Equation may be used to express the wave vector (kSP) of the propagated SPP perpendicular to a connection.

$$k_{SP}(\omega) = \frac{\omega}{c}\sqrt{\frac{\varepsilon_m(\omega)\varepsilon_d(\omega)}{\varepsilon_m(\omega)+\varepsilon_d(\omega)}}$$

where; *m* and *d*, respectively, stand for the metallic and dielectric materials' relative permittivity.

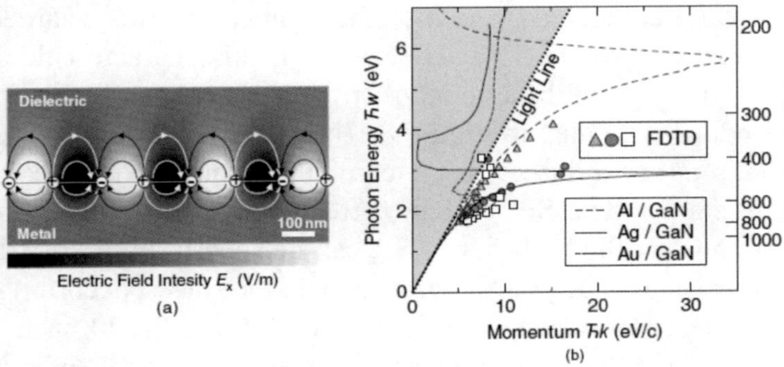

Figure 8.1. (a) The SPP propagation mode's electric field distribution was estimated using FDTD simulations. (b) SPP propagating mode dispersion diagram.

Source: https://www.researchgate.net/figure/a-SPP-propagation-lengths-at-1550-nm-measured-by-LRM-for-waveguide-with-one-two-and_fig4_258999877.

The phase velocity, as well as the speed of light in a medium, are, approximately, ω and C (Gontijot et al., 1999). The initial conditions of Maxwell's formula led to the development of this equation, which is known as the dispersal formula of the SP. The link is seen in the dispersion diagram. *k*SP and ω were discovered using equation and are shown in Figure 8.1(b) for the contacts Al/GaN, Ag/GaN, and Au/GaN, accordingly. Without regard to wavelength distribution, GaN's permeability was determined to be 6. The angular frequency (ω) the wave vector, and (k) can become photon energy ($\hbar\omega$) besides the momentum ($\hbar k$) through increasing by the Dirac constant ($\hbar$). The bright line provided is the black line.

$$k_d = \sqrt{\varepsilon_d \omega} / c$$

wherever; k_d seems to be the free space wavelength of both the light wave transmitted within the dielectric layer. It can only occur within the sheltered area of Figure 8.1(b), which is within the area where this line exists (Neogi et al., 2002; Okamoto et al., 2004).

However, since kSP might be greater than 1, the SP dispersal persists from outside the light line *k*d when $\varepsilon m + \varepsilon d < 0$. Given that this is the main crucial SP feature, producing SPPs requires a negative material permeability value. The big kSP indicates that even at the surface of the metal, the wavelengths were compacted, and also the momentum was retarded. The SPP may spread into smaller spaces because of this characteristic. Above eqn. indicated that even if the amount of $\varepsilon m + \varepsilon d$ is near 0, kSP gets infinite. The SPP dispersal diagrams in Figure 8.1(b), however, exhibit peak value since a metal's permeability is just a complex quantity with the formula $\varepsilon m = \varepsilon m' + \varepsilon m'' i$. The SP resonance frequency (ωSP), also known as that of the amount at the highest of kSP, can be seen in Figure 8.1(b) as a dotted line. For Al/GaN, Ag/GaN, with Au/GaN interfaces, accordingly, the SP resonating value (wavelength) is provided as 5.7 eV (220 nm), 2.9 eV (430 nm), and 2.3 eV (540 nm). This implies that Al, Ag, and Au have to be acceptable for usage in the corresponding UV, red, and blue zones (Xu et al., 2013).

8.3. HIGH-EFFICIENCY LIGHT EMISSIONS USING PLASMONICS

The creation of very brilliant light-emitting equipment is one prospective use of plasmonics. It has been predicted that new lighting sources would ultimately take the place of conventional fluorescent bulbs. For instance, quantum wells (QWs) based on InGaN provide powerful UV, blue, greenish, red, and white light emitters (Okamoto Kawakami, 2009). Additionally, organic LEDs that generate visible light are substantially less expensive than semiconductors and offer reliable lighting effects for screens and lighting sources. However, compared to fluorescent lights, their efficiency is still much lower. The improvement of emission efficiency is by far the most crucial demand for these organic and inorganic LEDs. Another of the best ways to raise these efficiency seems to be the SP coupled approach. Since its inception in 1990, the concept of SP-enhanced emitted light is often used to improve the emissions efficiency gains of several materials, including InGaN QWs. Using the setup depicted in Figure 8.2(a), Gontijo et al. revealed pairing the emissions from the InGaN QW into the SPP on silver thin film (Neal et al., 2005).

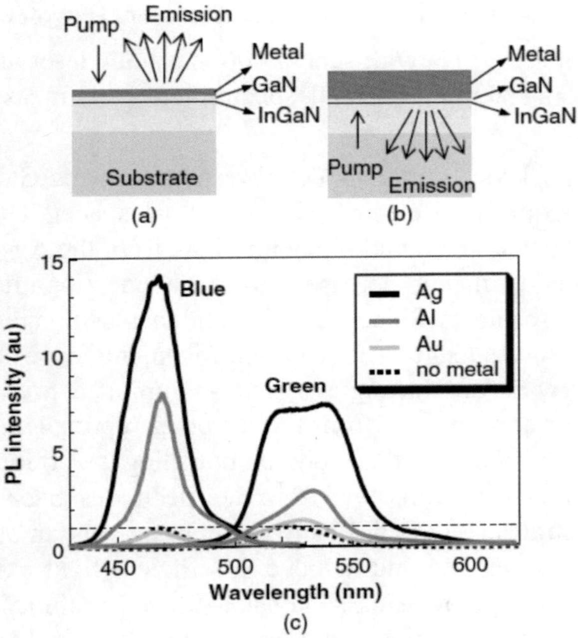

Figure 8.2. Sample design and experiment setup from prior research were provided.

Source: https://revisesociology.com/2017/11/26/the-steps-of-quantitative-research/.

However, they discovered that along with the SP connection, the PL intensity drastically dropped. Neogi et al. verified using that to ensure that respondents that now the time-resolved PL measurements could greatly increase the recombination process in such an InGaN/GaN QW. The SP connection was assumed to be detrimental to light-emitting substances within those early investigations because light could not have been recovered from the metal substrate effectively (Yeh et al., 2007).

We previously reported earlier than employing the arrangement indicated in Figure 8.3(b), that substantial photoluminescence (PL) increases from InGaN/GaN QW material covered using metal layers were seen. By using a metal-organic chemical deposition process, InGaN/GaN quantum well (QW) wafers were produced on a (0001) orientated sapphire substrate (MOCVD). A GaN (4 m) buffering layer, an InGaN QW (3 nm), and a GaN cap layer make up the QW heterojunctions (10 nm). By using high vacuum (HV) thermally vaporization, thin coatings of silvery, aluminum, or golden (50 nm) were placed on top of the surface of such wafers (Kwon et al., 2008).

By stimulating the QW with such a 406 nm diode laser and measuring the emission using a multi-channel spectrometer, PL measurements are carried out.

Figure 8.3 displays a typical Spectral response from a GaN/InGaNQW that is isolated from metal electrodes by a 10 nm spacer. The Absorption spectrum of either the green and blue emissions from the Ag- or Al-coated materials showed significant increases. The SP connection must be blamed for these PL improvements. Since Au's well-known plasmon resonating only happens at higher wavelength, really no improvements were achieved from materials coated with Au. The angle of photo-pumping, as well as detecting, is different in our arrangement from that of other reports (Lu et al., 2001; Feng and Okamoto, 2005). In this work, by polishing the bottom surface, we photo-pump and measure emissions from the specimens' backsides through to the clear substrates (Figure 8.3(b)). We could prevent an absorbing loss just at the metallic surface and achieve an efficient light extract through SPP just at the interface by using such back-side accessibility to QWs. So, we can employ extremely thick metal layers. It really should be taken into consideration to achieve greatly increased light output. Other SPP modes are formed just at air/metal contact of the opposite end of both the metal layer if indeed the metal layer is thinner than that of the depths of SPP penetrating. These SPP modes link with one another to create a symmetrical system that can be categorized as SPP modes. The SP frequencies and couplings conditions need to be altered as a result, making the photon extraction extremely challenging (Atwater and Polman, 2010).

Another benefit of the thick metal covering is that it prevents the silver's surfaces from oxidizing. The SP phase and wide range are altered by metal oxidation. But often, the oxidizing only occurs at the metal/air contact just not at the metal/semiconductor interaction. In this work, SPPs just at the material interface aid in the amplification of emitted light. The width of the metallic materials (50 nm) is sufficient to guarantee that now the metal/semiconductor interaction is not impacted by metal oxidizing just at the air/metal interface. Although it is a pretty straightforward method, the key technique that allowed us to initially gain mild improvements via the SP couplings is the back-side access (Rand et al., 2004).

8.4. THE MECHANISM FOR THE SP COUPLED EMISSIONS

We suggest a potential process for the light extraction of that QW-SP connection seen in Figure 8.3 (Kim et al., 2008). Whenever the energy of electron-hole pairs as well as the SP frequencies is comparable, electron-hole couples in the QW couple to plasma oscillation of electrons just at metal/semiconductor contact. Thus, a photon or phonon may well be replaced by an SPP as a result of an electron-hole recombine, increasing the recombination process and inner quantum efficiency. The SPP energy would've been thermally lost if the surface were flat. The SPP energy may be converted into light by adding roughness or microstructures to the atomic layers. High momentum SPPs may scatter, decelerate, and pair to radiative photons due to certain roughness. The very few tens of nanometer-sized nanostructures at the metallic surface were produced by adjusting the evaporating conditions throughout to achieve high photons extraction efficiencies (Zhang et al., 2017; Okamoto, 2021).

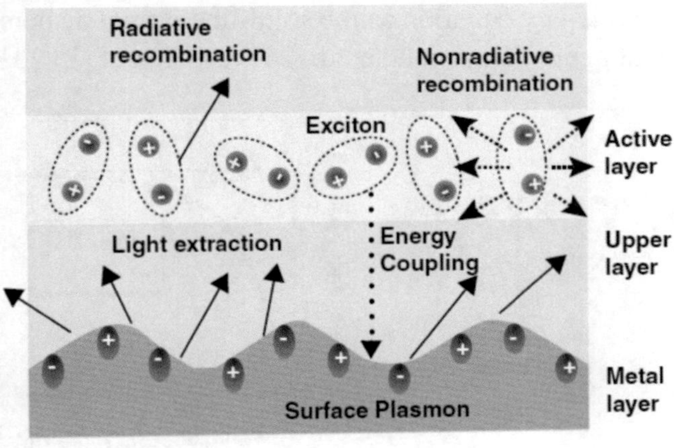

Figure 8.3. Schematic representation of the exciton-SP coupling's improved light emission effectiveness.

Source: https://www.researchgate.net/figure/Color-online-Schematic-representation-of-the-three-level-excitonic-system-together-with_fig20_233935957.

We used the 3D-FDTD simulation to assess the SP coupled method that we suggested (Huang et al., 2020; Jhuang et al., 2021). Poynting was employed to do the 3D-FDTD simulations. The predicted spatial pattern of an electromagnetic field near the metal/semiconductor contact is seen in Figure 8.4(a). The point source of light just at the connection caused the clear SPP state to emerge and spread across the interface. A pointed light source that makes do with an electronic configuration was a polarization plane waveform with a 525 nm wavelength with 1 V/m intensity. According to this finding, the SPP mode may be readily produced without the need for any specific structures through the straight transfer of energy between electron-hole pairs. Typically, specific setups like a grating coupling or perhaps attenuated total settings are required to produce SPP modes to fulfill a need for phase-matching among SPPs and rays. Nevertheless, irrespective of a phase matching requirement, the SPP mode may be created if indeed the light source is situated close to the material contact inside the wavelength range. This computation backs up the SP connectivity model we've suggested (Chen et al., 2017; Teng et al., 2018).

We computed the SPP phases at different wavelengths and depicted the SPP mode's polarizations in Figure 8.4 to validate the 3D-FDTD simulation. The lighting lines and SP frequencies, correspondingly, are shown by the dashed as well as dotted lines. The obtained results and the theoretical calculations provided by equation as the solid-line-drawn dispersion curves were in excellent agreement (Liang et al., 2014; Zhang et al., 2019).

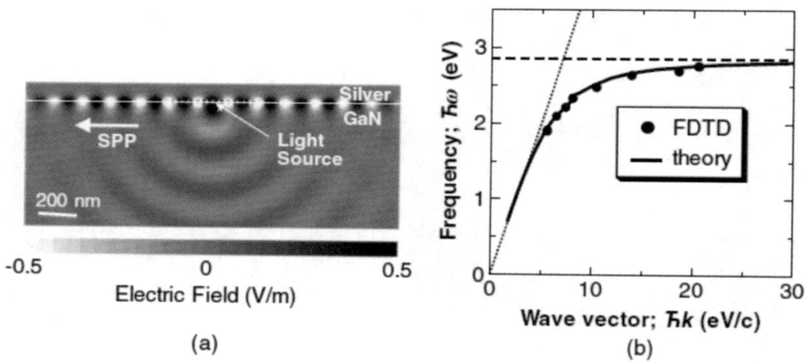

Figure 8.4. Calculated using the 3D-FDTD simulation, the generation, and propagation of SPP modes from the point light source at the silver/GaN interface.

Source: https://www.researchgate.net/figure/a-3D-FDTD-calculation-of-SPP-intensity-propagation-b-solid-line-Experimental_fig3_224541513.

We have gathered a lot of information that points to the possibility of the SP connection:

- We discovered that when GaN separator thickness is increased, the improved PL intensities drop rapidly. Given that the SPP is tightly constrained at the connection, this must make sense.
- We discovered that now the dispersal curves of the SPPs estimated using the dielectric characteristics and the observed wavelengths dependency of the PL increased ratios are strongly connected.
- By monitoring the temperature dependency of the PL intensity, we discovered that now the quantum efficiency energies were boosted.
- We also discovered that the moment PL measurements significantly boosted the spontaneously exhaust emissions. These data provide credence to the relationships that occur we suggest.

8.5. APPLICATIONS FOR ORGANIC MATERIALS

As previously mentioned, inorganic materials be the focus of the development of high-efficiency emission levels employing plasmonics. The fact that the SP couplings technology may be used with a variety of materials in addition to InGaN-based ones is its most significant benefit. We have thus attempted to apply this method to several natural light-emitting substances. For instance, steady laser beams for screens and illuminating sources are provided by polymers that have been suitably doped using molecules that emit visible light at such a price that is much less than that of inorganic materials.

Because they are lower cost and offer a variety of structural options available, organically LEDs are becoming commercially accessible and are utilized to replace inorganic LEDs. The path to spectrally wide white light polymeric emitters still faces several design obstacles despite the enormous potential of efficient solid-state illumination provided by these natural light emitters. Therefore, increasing the IQEs of both the polymeric dyes used for such light transmitters and the efficiency of light extract from certain organic films are both of economic and scientific significance. Here, we concentrate on exploiting the SP coupling to increase the light emitting efficiency of organic thin films (Liu and Nunzi, 2012; Liang et al., 2014).

The experimental design that we used to gauge various samples. By combining typical Coumarin 460 laser dyes in chlorobenzene, a dye polymeric solution is available. With UV stimulation, this laser dye produces blue light around 460 nm. To create 20 mM/L solutions of dye-doped polymers, 2% polymethylmethacrylate (PMMA) was added to the reaction mixture as just a host material. To do a quick comparison comparing polymeric emitters on top of metallization and polymer placed on quartz, just half of every substrate was metalized. The dye-doped PMMA films were spun onto the other silver and gold substrate following the plating procedure to yield a provision of related of around ~200 nm (Gu et al., 2011; Irandoost and Soleimani-Amiri, 2020). Figure 8.4 displays typical Coumarin 460 PL spectra on Ag, Au, and a substrate made of raw quartz.

The surface states could not appear to pair towards the emission spectra of Coumarin 460 to provide any discernible amplification, despite the Au's assistance in reflecting the pumping laser. However, owing to an interaction of plasmonic created just on Ag film and the fact that the plasmon resonant frequency roughly matched the dye's emissions frequency, we do notice an 11-fold augmentation of the emissions light from the Coumarin doped PMMA onto silver. The Coumarin 460 emissions wavelengths, as well as the dielectric properties for Ag, match rather well, especially if the results are normalized to 1. While reflections may contribute to part of the enhanced brightness, just the SP connection could explain the observed augmentation (Van Le et al., 2018).

For three conducting polymers (PF)-cyanophenylene (CNP) (1:1), PF-CNP (3:1), with polyfluorenes (PF)-triphenylamine (TPA)-quinoline (Q)—we also observed observable improvements in both PL intensity and exhaust emissions. These polymers have been employed as very effective emitted photons elements for organic LEDs.

8.6. DEVICE APPLICATION FOR LIGHT-EMITTING DEVICES

Figure 8.4 demonstrates the SP coupled and laser emission's power conversion system. The mechanisms of SP-exciton and SP-photon couplings provide novel channels for emission. If somehow the new emission route via the SP couplings is just much quicker than the previous emissions path [(2) + (11) > (3), (4) > (12)], therefore the highly efficient LEDs must be feasible. So, because SP-exciton, as well as SP-photon linking mechanisms, are reversible, plasmonics must also be capable of enhancing high-efficiency photovoltaic energy. Just at metal/dielectric interaction, the sunlight may be coupled to an SP and produce excitons within conducting polymers. If (8) > (7), (9) > (12) in Figure 8.4, then this procedure should enhance the efficiency of transmittance and photon energy conversions (Smythe et al., 2007; Hlubina et al., 2014).

In Figure 8.4, potential plasmonic LED device architectures are shown. The simplest form is Type A, which uses the typical LED design with such a p–n junction. Both electrical contacts and plasmon excitation may take place within a metallic surface (Damos et al., 2006; Bhatia and Gupta, 2011). The crucial aspect of this construction is that to have a strong SP coupling, the length between both the surface of the material as well as the InGaNQW must be extremely small. The p-type GaN coating must thus be less than 10 nm thick. As shown above, as the GaN spacer layer's thickness increases, the PL amplification ratios gradually decline. The element of SP couplings is quite challenging due to this property (Neff et al., 2006; Singh et al., 2013). We have previously made the Type A design; however, we were unable to significantly increase the emissions. There seem to be two causes. The first is because p-doping a GaN layer that was 10 nm thick was highly challenging. Secondly, the p-GaN layer is just too thin for us to get a satisfactory ohmic contact. The design classified as Type B contains metallic nanoparticles or metal nanoparticles just on wafer tops. This structure is likewise quite straightforward, and several organizations have created and reported on it. One InGaN/GaN single-QW LED design with the SP coupled effect was disclosed by Yeh et al. Between both the metal substrate as well as the InGaN QW-layer is a 70 nm p-type GaN layer, which is separated from the surface of the metal by a 10 nm p-type AlGaN present block layer. To produce an efficient SP connection, the overall length of 80 nm is just too great (Fujikawa et al., 2008; Chen et al., 2020).

Because of this, they were only able to increase the emissions by 1.5 times. A plasmonic LED with a similar structure to Type C was also described by Kwon et al. They initially added silver particles to the InGaN QW surface before adding an excessively thick GaN layer on top of the Ag particles. Only 3% of original Ag particles were left after the maximum temperature of a crystal formation vaporized the majority of them. As a result, they simply succeeded in increasing the emissions by 1.3 times. The

Type D device construction was created by Lu et al. utilizing dry etching and E-beam lithography (Fen and Yunus, 2011). The p-GaN layer's original diameter was 150 nm, as well as the etching depths, were 110 nm. As a result, the QW and the metal contacts were separated by 40 nm, which should be sufficient to establish the efficient exciton-SP coupling. With PL measures, this construction was able to achieve an improvement of 2.8 times; nevertheless, electrically all pumping of such a structure proved challenging but has not yet been successful. These minuscule amplification ratios shouldn't be appropriate for use in devices. As a result, a plasmonics-based extremely efficient LED architecture has still not been developed (Mahmudin et al., 2015).

As previously mentioned, technological challenges have prevented the development of efficient plasmonic LEDs. However, since they are so simple to make, organic circuits need to be more appropriate for use in plasmonic LED device topologies. Feng et al. demonstrated boosts of electroluminescence (EL) by employing organic matter in 2005, which would be the year after our initial study for the plasmonics-based PL enhancement. But on diffraction architecture of such an indium tin oxide (ITO)/quarts substrates, two layers of 80 nm thick tris-(8-hydroxyquinoline) aluminum (Alq) and 80 nm thick N, N-diphenyl-N, N-bis(1-naphthyl)-N, N-diphenyl-N, N-bis(1,1'-biphenyl)-4,4'-diamine) Increasing light extraction yield, the measured fluorescence intensity was four times more than that of un corrugated segment Includes. Using coupling with localized SP phases in such a thin line of Au nanoparticles within composite thin LEDs, Fujiki et al. were able to significantly boost EL in 2010. By easily altering the width of both the electron transport layer, such a structure, which has size-controlled Au nanoparticles implanted on ITO, may be utilized to achieve the ideal range for excited-state interaction. Comparing the molecular fluorescence to something like a traditional diode design, a 20-fold rise was seen. In addition, we successfully produced a significant PL increase for silicon nanoparticles in a silicon dioxide medium. Such indirect semiconductors typically have poor emissions efficiencies but using the SP pairing, it is feasible to raise these efficiency gains to levels comparable to straight compounded semiconductors. We think that extraordinarily brilliant silicon-based light-emitting systems might be produced using the SP linking technology (Figure 8.5) (Barakat et al., 2008; Srivastava and Gupta, 2011).

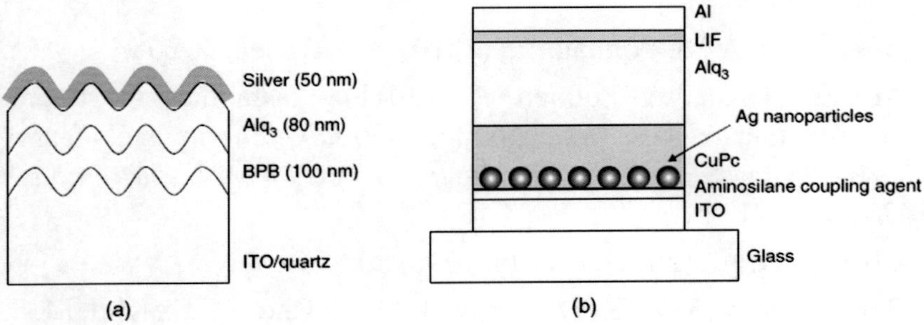

Figure 8.5. Reported electrically pumped plasmonic organic LEDs.

Source: https://pubs.acs.org/doi/10.1021/acsphotonics.1c01300.

REFERENCES

1. Atwater, H. A., & Polman, A., (2010). *Nat. Mater., 9*, 205.
2. Atwater, H. A., & Polman, A., (2011). Plasmonics for improved photovoltaic devices. *Materials for Sustainable Energy: A Collection of Peer-Reviewed Research and Review Articles from Nature Publishing Group,* 1–11.
3. Atwater, H. A., (2007). *Sci. Am., 296*, 56.
4. Barakat, N. A., Woo, K. D., Kanjwal, M. A., Choi, K. E., Khil, M. S., & Kim, H. Y., (2008). Surface plasmon resonances, optical properties, and electrical conductivity thermal hystersis of silver nanofibers produced by the electrospinning technique. *Langmuir, 24*(20), 11982–11987.
5. Barnes, W. L., Dereux, A., & Ebbesen, T. W., (2003). Surface plasmon subwavelength optics. *Nature, 424*(6950), 824–830.
6. Bhatia, P., & Gupta, B. D., (2011). Surface-plasmon-resonance-based fiber-optic refractive index sensor: Sensitivity enhancement. *Applied Optics, 50*(14), 2032–2036.
7. Catchpole, K. R., & Pillai, S., (2006). *J. Lumin., 121*, 315.
8. Chen, P., Xiong, Z., Wu, X., Shao, M., Meng, Y., Xiong, Z. H., & Gao, C., (2017). Nearly 100% efficiency enhancement of $CH_3NH_3PbBr_3$ perovskite light-emitting diodes by utilizing plasmonic Au nanoparticles. *The Journal of Physical Chemistry Letters, 8*(17), 3961–3969.
9. Chen, Y., Fang, L., Yi, D., Li, X., & Hong, X., (2020). Thermo-optic property measurement using surface plasmon resonance-based fiber optic sensor. *IEEE Sensors Journal, 20*(19), 11357–11363.
10. Damos, F. S., Luz, R. C., & Kubota, L. T., (2006). Investigations of ultrathin polypyrrole films: Formation and effects of doping/dedoping processes on its optical properties by electrochemical surface plasmon resonance (ESPR). *Electrochimica Acta, 51*(7), 1304–1312.
11. Derkacs, D., Chen, W. V., Matheu, P. M., et al., (2008). *Appl. Phys. Lett., 93*, 091107.
12. Derkacs, D., Lim, S. H., Matheu, P., et al., (2006). *Appl. Phys. Lett., 89*, 093103.
13. Fen, Y. W., & Yunus, W. M. M., (2011). Characterization of the optical properties of heavy metal ions using surface plasmon resonance technique. *Opt. Photonics J., 1*(03), 116–123.

14. Feng, J., & Okamoto, T., (2005). *Opt. Lett., 30*, 2302.
15. Ferry, V. E., Sweatlock, L. A., Pacifici, D., & Atwater, H. A., (2008). *Nano Lett., 8*, 4391.
16. Ferry, V. E., Verschuuren, M. A., Van, L. M. C., et al., (2011). *Nano Lett., 11*, 4239.
17. Fujikawa, R., Baryshev, A. V., Kim, J., Uchida, H., & Inoue, M., (2008). Contribution of the surface plasmon resonance to optical and magneto-optical properties of a Bi: YIG-Au nanostructure. *Journal of Applied Physics, 103*(7), 07D301.
18. Fujiki, A., Uemura, T., Zettsu, N., et al., (2010). *Appl. Phys. Lett., 96*, 3.
19. Gontijo, I., Boroditsky, M., Yablonovitch, E., et al., (1999). *Phys. Rev. B, 60*, 11564.
20. Gu, X., Qiu, T., Zhang, W., & Chu, P. K., (2011). Light-emitting diodes enhanced by localized surface plasmon resonance. *Nanoscale Research Letters, 6*(1), 1–12.
21. Hayashi, S., Kozaru, K., & Yamamoto, K., (1991). *Solid State Commun., 79*, 763.
22. Hlubina, P., Kadulova, M., Ciprian, D., & Sobota, J., (2014). Reflection-based fiber-optic refractive index sensor using surface plasmon resonance. *Journal of the European Optical Society-Rapid Publications, 9*.
23. Huang, C. L., Kumar, G., Sharma, G. D., & Chen, F. C., (2020). Plasmonic effects of copper nanoparticles in polymer photovoltaic devices for outdoor and indoor applications. *Applied Physics Letters, 116*(25), 253302.
24. Irandoost, R., & Soleimani-Amiri, S., (2020). Design and analysis of high efficiency perovskite solar cell with ZnO nanorods and plasmonic nanoparticles. *Optik, 202*, 163598.
25. Jhuang, L. S., Kumar, G., & Chen, F. C., (2021). Localized surface plasmon resonance of copper nanoparticles improves the performance of quasi-two-dimensional perovskite light-emitting diodes. *Dyes and Pigments, 188*, 109204.
26. Kakavelakis, G., Vangelidis, I., Heuer-Jungemann, A., Kanaras, A. G., Lidorikis, E., Stratakis, E., & Kymakis, E., (2016). Plasmonic backscattering effect in high-efficient organic photovoltaic devices. *Advanced Energy Materials, 6*(2), 1501640.

27. Kim, S. S., Na, S. I., Jo, J., et al., (2008). *Appl. Phys. Lett., 93*, 073307.
28. Kume, T., Hayashi, S., Ohkuma, H., & Yamamoto, K., (1995). *Jpn. J. Appl. Phys., 34*, 6448.
29. Kwon, M. K., Kim, J. Y., Kim, B. H., et al., (2008). *Adv. Mater., 20*, 1253.
30. Liang, Z., Sun, J., Jiang, Y., Jiang, L., & Chen, X., (2014). Plasmonic enhanced optoelectronic devices. *Plasmonics, 9*(4), 859–866.
31. Liu, F., & Nunzi, J. M., (2012). Enhanced organic light emitting diode and solar cell performances using silver nano-clusters. *Organic Electronics, 13*(9), 1623–1632.
32. Lu, C. H., Lan, C. C., Lai, Y. L., et al., (2001). *Adv. Funct. Mater., 21*, 4719.
33. Mahmudin, L., Suharyadi, E., Utomo, A. B. S., & Abraha, K., (2015). Optical properties of silver nanoparticles for surface plasmon resonance (SPR)-based biosensor applications. *Journal of Modern Physics, 6*(08), 1071.
34. Mapel, J. K., Singh, M., Baldo, M. A., & Celebi, K., (2007). *Appl. Phys. Lett., 90*, 121102.
35. Nakamura, S., Senoh, M., & Mukai, T., (1993). *Jpn. J. Appl. Phys. Part 2 - Lett. Express Lett., 32*, L8.
36. Nakayama, K., Tanabe, K., & Atwater, H. A., (2008). *Appl. Phys. Lett., 93*, 121904.
37. Neal, T. D., Okamoto, K., & Scherer, A., (2005). *Opt. Express, 13*, 5522.
38. Neal, T. D., Okamoto, K., Scherer, A., et al., (2006). *Appl. Phys. Lett., 89*, 221106.
39. Neff, H., Zong, W., Lima, A. M. N., Borre, M., & Holzhüter, G., (2006). Optical properties and instrumental performance of thin gold films near the surface plasmon resonance. *Thin Solid Films, 496*(2), 688–697.
40. Neogi, A., Lee, C. W., Everitt, H. O., et al., (2002). *Phys. Rev. B, 66*, 153305.
41. Okamoto, K., & Kawakami, Y., (2009). *IEEE J. Sel. Top. Quantum Electron., 15*, 1199.
42. Okamoto, K., & Kawakami, Y., (2010). *Phys. Status Solidi C, 7*, 2582.

43. Okamoto, K., (2021). Plasmonics for light-emitting and photovoltaic devices. *Organic Semiconductors for Optoelectronics,* 343–357.
44. Okamoto, K., Niki, I., Scherer, A., et al., (2005). *Appl. Phys. Lett., 87,* 071102.
45. Okamoto, K., Niki, I., Shvartser, A., et al., (2004). *Nat. Mater, 3,* 601.
46. Pryce, I. M., Koleske, D. D., Fischer, A. J., & Atwater, H. A., (2010). *Appl. Phys. Lett.* (pp. 1-25)
47. Rand, B. P., Peumans, P., & Forrest, S. R., (2004). *J. Appl. Phys., 96,* 7519.
48. Singh, S., Mishra, S. K., & Gupta, B. D., (2013). Sensitivity enhancement of a surface plasmon resonance based fiber optic refractive index sensor utilizing an additional layer of oxides. *Sensors and Actuators A: Physical, 193,* 136–140.
49. Smythe, E. J., Cubukcu, E., & Capasso, F., (2007). Optical properties of surface plasmon resonances of coupled metallic nanorods. *Optics Express, 15*(12), 7439–7447.
50. Srivastava, S. K., & Gupta, B. D., (2011). Influence of ions on the surface plasmon resonance spectrum of a fiber optic refractive index sensor. *Sensors and Actuators B: Chemical, 156*(2), 559–562.
51. Stuart, H. R., & Hall, D. G., (1996). *Appl. Phys. Lett., 69,* 2327.
52. Teng, N. W., Yang, S. S., & Chen, F. C., (2018). Plasmonic-enhanced organic photovoltaic devices for low-power light applications. *IEEE Journal of Photovoltaics, 8*(3), 752–756.
53. Van Le, Q., Jang, H. W., & Kim, S. Y., (2018). Recent advances toward high-efficiency halide perovskite light-emitting diodes: Review and perspective. *Small Methods, 2*(10), 1700419.
54. Xu, X., Funato, M., Kawakami, Y., et al., (2013). *Opt. Express, 21,* 3145.
55. Yeh, D. M., Huang, C. F., Chen, C. Y., et al., (2007). *Appl. Phys. Lett., 91,* 171103.
56. Zhang, X., Xu, B., Wang, W., Liu, S., Zheng, Y., Chen, S., & Sun, X. W., (2017). Plasmonic perovskite light-emitting diodes based on the Ag–$CsPbBr_3$ system. *ACS Applied Materials & Interfaces, 9*(5), 4926–4931.

57. Zhang, Y., Sun, H., Zhang, S., Li, S., Wang, X., Zhang, X., & Guo, Z., (2019). Enhancing luminescence in all-inorganic perovskite surface plasmon light-emitting diode by incorporating Au-Ag alloy nanoparticle. *Optical Materials, 89*, 563–567.

INDEX

Printed in the United States
by Baker & Taylor Publisher Services